Orang-utans

WWF *for a living planet*®

Orang-utans

Behaviour, Ecology and Conservation

JUNAIDI PAYNE AND CEDE PRUDENTE

With every purchase of *Orang-utans: Behaviour, Ecology and Conservation*, New Holland Publishers (UK) will donate 1% of the net price to support WWF's conservation goals.

Learn more about WWF at www.worldwildlife.org

First published in 2008 by New Holland Publishers (UK) Ltd
London • Cape Town • Sydney • Auckland
www.newhollandpublishers.com

10 9 8 7 6 5 4 3 2 1

Garfield House, 86–88 Edgware Road, London W2 2EA, UK
80 McKenzie Street, Cape Town 8001, South Africa
Unit 1, 66 Gibbes Street, Chatswood, NSW 2067, Australia
218 Lake Road, Northcote, Auckland, New Zealand

ISBN: 978 1 84537 928 5

Editorial Director: Jo Hemmings
Consultant Editor: James Parry
Project Editor: Kate Parker
Design and cover design: Studio Ink (info@studioink.co.uk)
Cartography: William Smuts
Production: Melanie Dowland

Reproduction by Pica Digital PTE Ltd, Singapore
Printed and bound by Tien Wah Press, Singapore

Note: The author and publishers have made every effort to ensure that the information given in this book is safe and accurate, but they cannot accept liability for any resulting injury or loss or damage to either property or person, whether direct or consequential and howsoever arising.

Front cover *Orang-utan – red ape of the Borneo and Sumatra rainforests;*
Back cover *Infant at Sepilok, Sabah;* **Page one** *Adult female Bornean orang-utan, Tanjung Puting National Park, Central Kalimantan, Indonesia;*
Page two *Adult female Sumatran orang-utan, Gunung Leuser National Park, North Sumatra, Indonesia;* **Opposite** *Female and infant Sumatran orang-utan, Gunung Leuser National Park, North Sumatra, Indonesia.*

CONTENTS

Opposite and right (top) *Awesome faces of wild orang-utans, glimpsed in the tropical rainforests of Borneo and Sumatra.*

FOREWORD

In one of the last interviews before his death in 2005, HRH Prince Bernhard of the Netherlands, WWF's first president, expressed his concern for the future of Borneo's forest habitat, home to the threatened orang-utan.

Looking into the eyes of an orphaned baby orang-utan, he said, is a look you'll never forget.

These words remind me of the passion Prince Bernhard and the founders of WWF had for wildlife and conservation, a passion that four decades later continues to be central to the work that we do.

From the halls of government to the grassroots level, we are working with partners to secure a future for the orang-utan. This requires a concerted effort to not only provide protection for the threatened species, but also to fight against the destruction of its rich habitat by unsustainable logging, forest fires and forest conversion for agriculture.

An historic tri-country declaration signed between Brunei Darussalam, Indonesia and Malaysia commits those governments to conserving and sustainably managing 22 million hectares of equatorial rainforests that are the heart of Borneo, and a lifeline for the people and wildlife who live there. Three countries sharing that one conservation vision gives me hope that one of the most important areas for biodiversity in the world, and home to the orang-utan, can be protected. Let us hope that a vision of similar magnitude can be formulated within Indonesia for that other great equatorial island of Sumatra, where orang-utans remain endangered.

This well-documented, in-depth book by Junaidi Payne and Cede Prudente is an essential resource for those who want to gain a deeper understanding of one of the world's most intelligent species and learn more about the conservation efforts dedicated to improving its long-term survival.

As the book highlights, there are viable solutions that can help safeguard and restore critical orang-utan habitat. Governments, individuals, and businesses operating in the region all have a role to play.

We at WWF hope that readers everywhere will heed the author's call to action, and join us in our mission to build a future where we can live in harmony with these magnificent creatures.

James P. Leape, Director General
WWF International

Opposite *Adult female Sumatran orang-utan in strangling fig plant, Gunung Leuser National Park.*

CHAPTER ONE

A PORTRAIT OF THE SPECIES

THE ORANG-UTAN IS A MEMBER OF THE ZOOLOGICAL FAMILY KNOWN AS THE GREAT APES, WHICH INCLUDES HUMANS, CHIMPANZEES AND GORILLAS. THE GREAT APES BELONG TO A MUCH BIGGER GROUPING KNOWN AS PRIMATES, WHICH INCLUDES LESSER APES, MONKEYS, LEMURS AND BUSH-BABIES. THE MAJOR DIFFERENCE BETWEEN AN APE AND A MONKEY IS THAT AN APE HAS NO TRACE OF A TAIL, WHILE MOST MONKEYS HAVE A TAIL, EVEN IF IT IS A MERE STUMP. TWO CHARACTERISTICS COMMON TO ALL PRIMATES ARE THAT THEY HAVE EYES ON THE FRONT OF THEIR HEAD, AND HANDS AND FINGERS WHICH ARE ABLE TO GRASP.

A VERY SPECIAL RED APE

Orang-utans occur wild only on two equatorial islands: Borneo (the third largest island, shared by the nations of Indonesia, Malaysia and Brunei) and Sumatra (Indonesia). Although closely related, the orang-utan populations on the two islands are now commonly regarded as two species: Bornean orang-utan (scientific name *Pongo pygmaeus*) and Sumatran orang-utan (*Pongo abelii*).

People are fascinated by orang-utans, perhaps because they seem so similar to us while being so very different; similar in their physical appearance and intelligence, yet different in that they live in an environment which is so foreign to us, up in the trees, without shelter, and often drenched in pouring equatorial rain, leading a quiet and often solitary existence in the tropical rainforests. Seen in zoos, orang-utans may remind us of clowns. In zoos and at rehabilitation centres in their native countries, orang-utans seem as much at home on the ground as in tree tops, and they show no fear of human beings. To us, infant orang-utans appear cute, pathetic, and similar to human babies in their tantrums, needs, inquisitiveness and facial expressions. Unable to

fend for themselves, young orang-utans seen close up remind us more of delinquent, hairy children than of wild animals. But truly wild orang-utans induce different feelings. Usually only glimpsed, as if from another world in the gloom, humidity and insect noises of the forest, they are nervous of humans. They are no larger than us, yet strong and awesome.

The exact coloration of an orang-utan depends on its age and ancestry. Young Sumatran orang-utans may be a pale orange-red, while old male Bornean orang-utans may be reddish-black. Infants have a brownish face with large, pale pink marks around their eyes and mouth. Once familiar with both Bornean and Sumatran orang-utans, we can usually distinguish them by observing several subtle characteristics. Most notably, Sumatran orang-utans usually have whitish hairs on the face and groin, unlike Bornean individuals.

Orang-utans do not normally stand up straight, so saying how tall they are does not make much sense. However, if an adult male were to stand stretched up, it would typically reach 1.4 metres from head to foot. The sitting-up height of baby orang-utans is between 30 and 40 centimetres, and of adults 70 to 90 centimetres. Orang-utans may seem to be smaller when

Previous pages *Adult male Bornean orang-utan in wild fig tree, Kinabatangan Wildlife Sanctuary, Sabah.*

Left *Infant orang-utans have a brownish face with large, pale pink marks around their eyes and mouth.*

Opposite *Wild orang-utans are no larger than humans – less than 90cm in height when sitting – but are physically very strong. Adult males have a total outstretched arm span of up to 2.4m. This Sumatran male is making a 'long call' (see page 73).*

seen at close quarters on the ground than when seen at a distance in the trees. One reason for this lies in the relative length of the legs; about 50 per cent of the height of a human being is accounted for by the legs, while in orang-utans the equivalent is about 40 per cent. More significant is the fact that the orang-utan's arm span is greater than its body height. At up to 2.4 metres when outstretched, the adult male's arms make him look enormous, especially up in the trees. Male orang-utans are the world's largest tree-dwelling mammals but, owing to their weight and bulk, they sometimes have to descend and travel on the ground, even in their rainforest habitat.

From birth to the age of about eight years, male and female orang-utans are similar in size and appearance, although males average a little larger. After this age males continue to grow, ending up much bigger than females. Wild mature females of ten years or more weigh an average of about 33 to 45 kilograms; wild mature males above 15 years are much bigger, weighing about 70 to 90 kilograms. Well cared-for zoo animals are usually larger and heavier than their wild counterparts: mature females in captivity may reach 80 kilograms, males 190 kilograms. Indeed, many zoo orang-utans become obese, a consequence of a daily high-quality diet and little exercise. All orang-utans are extremely strong for their size, a fact belied by the scrawny arms under their long hair.

Only mature males develop powerful shoulders, a throat pouch (used to make loud noises by expelling air) and wide tissue flanges on the cheeks. There exists a unique phenomenon, called bimaturism (see pages 102–103), among male orang-utans both in the wild and in zoos, whereby normally only one adult male in a particular locality develops full, massive body size with cheek flanges. Other male orang-utans in the vicinity, although adult and sexually mature, remain in a less-developed form until the big male dies or disappears.

These differences in body size between the two types of adult male, and between male and female, as well as differences in coloration due to origin and age, caused

Opposite *Young adult male orang-utan without cheek flanges, Kinabatangan, Sabah.*

Left *Mature male orang-utan with cheek flanges and throat pouch, Gunung Leuser, Sumatra.*

puzzlement amongst zoologists of the 19th century, who thought that the variations indicated the existence of several species. But the size differences are of some help to people who catch glimpses of wild orang-utans. A 'very large' wild orang-utan will be the full-sized male, a 'large' one is likely to be the less-developed male, a 'medium-sized' orang-utan may be a mature female (and the observer should try to see whether there is a small orang-utan nearby), while a solitary 'small' wild orang-utan is likely to be an adolescent that has recently left its mother.

Orang-utans have hands similar to those of other apes and monkeys. The feet are very similar to the hands, with long flexible toes, and the big toe (known scientifically as the hallux) is opposable to the other toes. The hip joint is as flexible as the shoulder and arm joint. Together, these features make the orang-utan well able to climb and travel through and between trees. Younger and lighter individuals are even able to hang upside down by their feet.

Both orang-utan species live only in natural tropical rainforest, feeding primarily on fruits, with smaller quantities of leaves, bark, flowers and invertebrate animals. They are active only during the daytime, and only very rarely descend from the tree canopy to the ground. In the late afternoon, every adult and near-mature wild orang-utan makes a simple nest for the night, within the crown of a tree, consisting of small branches snapped and bent towards each other to form a springy platform.

Individual female orang-utans give birth to only a few offspring during their lifetime, singly, and at intervals in excess of five years. Males play no role in caring for the young. Both in the wild and in zoos, orang-utans have a lifespan of over 30 years and many live to 50. Wild orang-utans are among the least sociable of all mammals, spending most of their life alone after they have left their mother. Groups of orang-utans living in close proximity in zoos are much more sociable, and generally amicable.

Top *Female and young adolescent orang-utan, Sepilok, Sabah. Individual female orang-utans give birth to only a few offspring during their lifetime, singly, and at intervals in excess of five years.*

Above *The orang-utan's foot is similar to its hand. The big toe (known as the hallux) is opposable to the other toes.*

Left *Orang-utans often behave in ways that may seem comical to humans – young ones like to play while orang-utans of all ages like to experiment.*

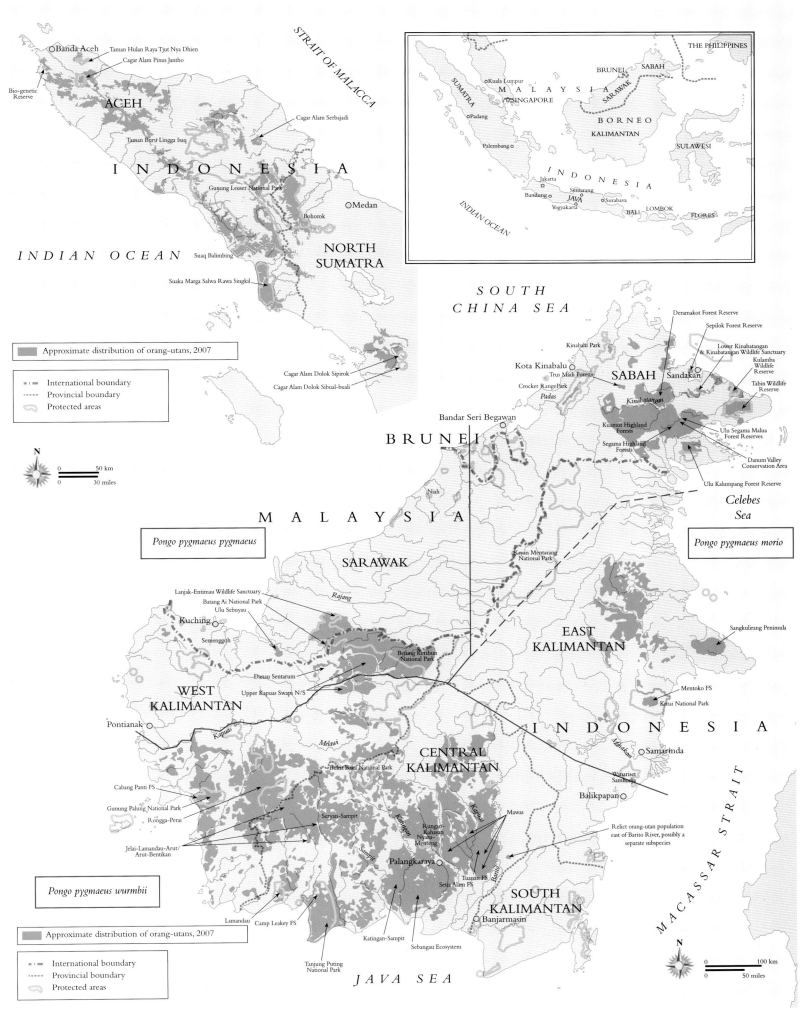

STRAIT OF MALACCA

Banda Aceh
Taman Hulan Raya Tjut Nya Dhien
Cagar Alam Pinus Jantho

Bio-genetic Reserve

ACEH

Cagar Alam Serbajadi

Taman Buru Lingga Isaq

I N D O N E S I A

THE PHILIPPINES

BRUNEI   SABAH

KUALA LUMPUR
MALAYSIA
SINGAPORE   SARAWAK

SUMATRA

BORNEO

KALIMANTAN

SULAWESI

Padang

Palembang

I N D O N E S I A

Jakarta
Bandung   Semarang
JAVA   Surabaya
Yogyakarta   Surabaya
BALI   LOMBOK

INDIAN OCEAN

FLORES

Gunung Leuser National Park

Bohorok

Medan

I N D I A N   O C E A N

NORTH SUMATRA

Suaq Balimbing

Suaka Marga Salwa Rawa Singkil

Cagar Alam Dolok Sipirok
Cagar Alam Dolok Sibual-buali

Approximate distribution of orang-utans, 2007

International boundary
Provincial boundary
Protected areas

N

0   50 km
0   30 miles

S O U T H
C H I N A   S E A

Deramakot Forest Reserve
Sepilok Forest Reserve

Kinabalu Park

Lower Kinabatangan & Kinabatangan Wildlife Sanctuary
Kulamba Wildlife Reserve

Kota Kinabalu
Trus Madi Forests
SABAH   Sandakan
Tabin Wildlife Reserve

Crocker Range Park
Padas
Kinabatangan

Bandar Seri Begawan

B R U N E I

Kuamut Highland Forests

Segama Highland Forests

Ulu Segama Malua Forest Reserves

Danum Valley Conservation Area

Ulu Kalumpang Forest Reserve

Celebes Sea

Niah

M A L A Y S I A

Pongo pygmaeus pygmaeus

SARAWAK

Kayan Mentarang National Park

Pongo pygmaeus morio

Lanjak-Entimau Wildlife Sanctuary
Batang Ai National Park
Ulu Sebuyau

Kuching
Semenggoh

EAST KALIMANTAN

Rajang

Betung Kerihun National Park

Sangkulirang Peninsula

Danau Sentarum
Upper Kapuas Swaps N/S

WEST KALIMANTAN

Mentoko FS
Kutai National Park

Pontianak

Kapuas

Melawi

I N D O N E S I A

Mahakam   Samarinda

Bukit Baka National Park

CENTRAL KALIMANTAN

Wanariset Sambodja

Cabang Panti FS

Gunung Palung National Park
Rongga-Perai

Seryan-Sampit

Mawas

Balikpapan

Jelai-Lamandau-Arut/
Arut-Bentikan

Katingan

Rungan-Kahayan Nyaru-Menteng

Kapuas

Relict orang-utan population east of Barito River, possibly a separate subspecies

Sampit

Palangkaraya

Barito

Pongo pygmaeus wurmbii

Lamandau   Camp Leakey FS

Katingan-Sampit

Tuanan FS
Setia Alam FS

SOUTH KALIMANTAN

Banjarmasin

M A C A S S A R   S T R A I T

Approximate distribution of orang-utans, 2007

International boundary
Provincial boundary
Protected areas

Tanjung Puting National Park

Sebangau Ecosystem

J A V A   S E A

N

0   100 km
0   50 miles

Based on Singleton et al, 2004 (see Further Reading)

CHAPTER TWO

TROPICAL RAINFORESTS – HOME OF THE ORANG-UTAN

THE BEAUTY AND COMPLEXITY OF MATURE TROPICAL RAINFOREST CAN BE HARD TO DESCRIBE ADEQUATELY; APPROPRIATE WORDS MIGHT INCLUDE TALL, HUMID, HOT, LUXURIANT, BEWILDERING, AWESOME, GRAND, SOMBRE, DIVERSE AND INTRICATE. AT FIRST GLANCE, A TROPICAL RAINFOREST SEEMS TO BE MERELY AN ENDLESS EXPANSE OF GREEN AND BROWN, LEAVES AND TREE TRUNKS, NOISY WITH INSECTS BUT BARELY A TRACE OF ACTIVITY. THIS VIEW IS DECEPTIVE.

Tropical evergreen rainforests are the natural vegetation of Borneo (about 740,000 square kilometres) and Sumatra (about 475,000 square kilometres), as well as the Sunda continental shelf, which includes Peninsular Malaysia, Java and Palawan (southern Philippines). Borneo and Sumatra have a wet equatorial climate, with daytime temperatures of around 27°C inside the lowland forests and over 30°C in exposed areas. Most areas receive between 1,800 and 5,000 millimetres of rain each year. Rain may fall at any time, but there tend to be periods with especially heavy and frequent rainfall, particularly from November to February and from May to August. Humidity is very high during rainy periods. There is a variety of soil types on both islands but three general points are evident, all of which have implications for the distribution and habits of orang-utans. First, the greater the inherent fertility of a particular soil type, the quicker plants can grow, and the more often and more abundantly those plants produce flowers and fruits, the latter being the main foodstuff of orang-utans. Second, owing to its volcanic origins, northern Sumatra has generally more fertile soils than Borneo. Third, the prevailing heavy rainfall throughout the region tends to leach nutrients downwards, both into the ground and downhill, so alluvial flat lowlands can be much more fertile and biologically productive than hills, mountains and sloping lands.

Tropical rainforests are home to plants of every conceivable size, shape, texture and hue. Trees represent the essence of any forest in terms of its structure. There are big trees, medium-sized trees, small trees, lianas (woody climbing plants), palms, shrubs, herbs, ferns, orchids, mosses and many more. The abundant water, sunlight and heat of the humid tropical zone ensure continuous growth. Countless forms of life, mostly small and hidden from view, are struggling to grow and reproduce themselves. Microscopically small fungi coat the roots of large trees, taking energy-giving molecules and in return helping the trees to extract scarce minerals from the soil. Termites eat wood and, in the depths of safe nest chambers, save their waste products on which to grow tiny mushrooms for food. Dung beetles seek animal droppings and longhorn beetles seek tall timber trees, both for the same reason – food for their young. Insects tiny and large pollinate flowers, while others seek rotting fruits on the forest floor. Trees die from the ravages of fungi and bacteria, and topple to leave gaps in the forest canopy for young plants to colonize and repeat the struggle of life. Borneo and Sumatra together sustain 20,000 or more species of plant, all but a few of them ultimately dependent for their survival on being part of the forest system. In Borneo and Sumatra, trees are a major element in characterizing the forest. Borneo alone has over 540 genera of trees, incorporating thousands of tree species.

PRIMARY OR OLD GROWTH WETLAND FORESTS

The plant species composition, forest tree structure and overall height of the forest canopy vary markedly from place to place, according to soil characteristics, slope and altitude above sea level. More than a dozen fairly distinct types of primary or old growth forest – those that have never or only minimally been disturbed by humans – can be recognized.

Starting from the coast, on sandy shorelines there is beach forest, often characterized by *Calophyllum inophyllum* (Borneo Mahogany) and Rhu trees (*Casuarina equisetifolia*, superficially resembling Christmas trees). Most of this forest has given way to human settlement. Orang-utans may occasionally enter a few remnant sites in eastern Sabah.

Much more extensive are mangrove forests, consisting of a few genera of trees which are specialized to grow on muddy coastlines periodically inundated with tidal seawater. All types of mangrove forest trees are characterized by having aerial or stilt roots, which protrude above the surface of the anaerobic mud as a means of absorbing oxygen. These forests are most extensive at, around and between the estuaries of large rivers. Orang-utans

sometimes enter mangrove forests where there is adjacent freshwater, peat swamp or lowland dipterocarp forest, the habitats that represent their normal range. In eastern Sabah, there are a few individual orang-utans, trapped and surviving in mangrove as long as 20 years after the adjacent inland forests were converted to plantations. Although these tiny orang-utan populations may have no future, they are instructive in demonstrating that individual orang-utans may survive for long periods in habitats that will not be able to support breeding populations in the long term.

Related to mangrove are nipa palm stands, which may grow extensively where freshwater flows into tidal mudflats. These areas are among the natural habitats most impenetrable to humans, but orang-utans enter them from their range in adjacent inland forests, to feed on the nipa seeds. In some places, transitional forests may be recognized where mangrove and nipa vegetation merge with peat and freshwater swamp forests. In parts of Borneo these forests seem to sustain high population densities of orang-utans, one reason being that they have been squeezed into these habitats following conversion of adjacent land to plantations since the 1980s. But it is also likely that the variety of vegetation in one place helps to ensure year-round sources of fruits and other orang-utan foods.

Progressing inland from the coastal swamps we find three broad types of forest habitat that have two things in common: they are waterlogged, or subject to periodic inundation with freshwater; and, in regions where orang-utans occur, these forests tend to be favourite habitats, with a higher population density (see page 86) of orang-utans than is found in nearby dry land forests.

Peat swamp forests occur on peaty soils, which developed on former mangrove areas or freshwater swamps in Borneo and Sumatra only within the past 10,000 years. The process of peat formation starts on open, acidic, muddy soils, conditions in which few organisms of decay can survive. Any trees and other plants present die, but do not decay, instead building up a soggy mass and forming an organic soil with little mineral component. Under equatorial tropical conditions, the process of peat build-up is fast, as there are ample water and sunlight but minimal decay. These soils and forests are therefore among the youngest in Borneo, but they exhibit great variation in tree species composition, mainly according to the depth of the peat layer. Tree pollen is found in 20-million-year-old Sarawak coal (fossilized peat), and the same types of tree are found in 4,000-year-old peat swamp forests existing today. This is a striking illustration that, despite human onslaughts on the Bornean forests in recent decades, nature and natural processes can bring repeated rebirths of ecosystems on geological time scales.

There are two fairly distinct types of peat formation. Ombrogenous peat builds up to depths of 20 metres or more, the central part of the system being the deepest and the outer areas shallowest. Eventually, the central parts of a peat swamp system get to be higher than the outer, drying out often, and supporting a simple forest type. Ombrogenous peat soils cover vast areas mainly on the south and west side of the island of Borneo and the east side of Sumatra. In central Kalimantan, these peat swamp forests support orang-utans at population

Opposite *Mangrove forests and nipa palm stands form the coastline in many parts of Borneo and Sumatra.*

Above *Riverside peat swamp forest, Sekunyir River, Central Kalimantan.*

densities of between one and four individuals per square kilometre. It is probable that, in the past, some peat swamp forests in Sarawak supported high densities of orang-utans. In the mid-19th century, the explorers Wallace and Hornaday readily obtained their many orang-utans shot for museums from peat swamp forests east of Kuching. Small remnant populations of orang-utans still exist in these swamp forests.

In contrast, topogenous peat swamps develop in freshwater swamp basins, are not as deep as ombrogenous peat soils, and usually have a mix of silt, so are more fertile. Topogenous peat soils tend to cover rather smaller areas on the east side of Borneo and the west coast of Sumatra. In the Kluet topogenous swamps on the west side of the Leuser ecosystem, studies have found orang-utan population densities of between three and seven individuals per square kilometre. Surveys in the Kinabatangan floodplain in Sabah in the mid-1980s, when the forests were still in good condition, indicated orang-utan population densities of about three individuals per square kilometre. These forests are mixed topogenous peat and freshwater swamp forest. More recent surveys of the best remaining forest in this region, after expansion of plantations through the adjacent dry lands since 1984, indicate population densities of three to six per square kilometre.

Freshwater swamp forests occur on low-lying mineral soils, where the water table is high, the land is periodically inundated with freshwater from rain and nearby rivers, and where no peat develops. Soils in these circumstances tend to be fertile. In some areas, this habitat merges with topogenous peat swamps. The exact species composition of trees varies markedly from place to place, in part influenced by how wet the soils are for most of the time, and part by chance events in the past. Usually, the more constantly waterlogged the soil, the lower is the forest canopy and the lower the diversity of tree species. The majority of these forests have now been converted to plantations or had most of their trees removed for timber.

Right *Freshwater swamp forests and oxbow lakes near Sukau, lower Kinabatangan, Sabah.*

The third wet forest type is lowland riverine or riparian forest, a strip of forest roughly 5 to 50 metres wide, along river banks. This differs from freshwater and peat swamp forest because the rivers frequently burst their banks after heavy rain, depositing a layer of fertile silt. Thus, the river bank zone consists of very fertile soil which is at a slightly higher level than both the normal river water and the swamps further away. Since the courses of the large lowland rivers in Borneo are in a process of constant slow change, the pattern and extent of riverine forest changes accordingly. Sometimes, the course of a large river cuts through a bend, forming an oxbow lake which retains its surrounding riverine forest, and then the lake fills up with silt, which in turn becomes tree-covered, thereby contributing to the diversity of forest types within a floodplain. Overall, it is probably a combination of the great variety of tree species composition, frequent wetness and soil fertility (except in ombrogenous peat), that lies behind the evident suitability of swamp and riverine forest for orang-utans. These combinations tend to result in near-constant availability of fruits and a great variety of fruit species. Wetland forests can be found in the extensive coastal plains of south and south-western Borneo, Kinabatangan in Sabah, and in Aceh, Sumatra.

PRIMARY OR OLD GROWTH DIPTEROCARP FORESTS

The natural vegetation on most non-swampy soils throughout Borneo and Sumatra, from sea level up to about 1,300 metres above sea level, is a mix, classified together as dipterocarp forest. Although the tree species composition of these forests varies greatly from region to region, and also locally, according to small differences in soil type and slope, they are all characterized by having a tall tree canopy (40 to 60 metres on average, with some 'emergent trees even taller), with the majority of large trees belonging to a single plant family, the Dipterocarpaceae. The enormous variety within dipterocarp forests makes it impossible to devise a single classification useful to everyone. Botanists, zoologists and foresters may have differing ideas, but all classifications are in some way linked to soil characteristics, altitude (which influences temperature and cloud cover) and the most prominent tree species. When thinking of orang-utans and other large mammals, one can best imagine four types of dipterocarp forest in Borneo and one type for Sumatra. For Borneo, all extreme lowland dipterocarp forests seem to be excellent habitat for orang-utans. At altitudes of less than

about 150 metres above sea level, and with flat to gentle topography and fertile moist soils, these forests tend to be somewhat less diverse in plant species than forests at higher elevations. In addition to dipterocarps, many of the tallest trees belong to the legume family; in some areas, there are numerous ironwood trees (*Eusideroxylon zwageri*), a species unique to Borneo, Sumatra and some intervening islands. Except for small remnants in a very few protected areas, mainly in eastern Sabah, these forests have either been almost entirely converted to plantations and agriculture, or they are badly damaged and few of the original trees remain. Where orang-utans are absent from this type of forest, it seems likely that prehistoric (for example, around Niah, Sarawak) or more recent (for example, much of Kalimantan) hunting pressure by native people has exterminated the species. Where orang-utans are present in this forest type, and in the absence of significant hunting, the usual population density of orang-utans is one to three individuals per square kilometre. Between roughly 150 and 300 metres above sea level is lowland dipterocarp forest (botanists do not distinguish between this and extreme lowland dipterocarp forest). Tree species composition varies greatly between regions and soils, but is normally very diverse, and there may be considerable local

variation in topography. In Lambir Hills National Park, Sarawak, a remnant of this forest type at 170 metres above sea level has the greatest known tree species diversity recorded in the Old World (Asia and Africa), with almost 1,200 tree species identified within a 53-hectare plot. In regions where the Bornean orang-utan occurs, it makes use of this forest type, but much less than the extreme lowland dipterocarp forest, with a population density of perhaps one orang-utan per two to three square kilometres.

Opposite *Old growth dipterocarp forests of Danum Valley Conservation Area, with logged forest in the foreground.*

Above *Old growth lowland dipterocarp forest, Imbak Canyon, Sabah, here dominated by red camphorwood trees.*

Left Hill dipterocarp forest, Leuser ecosystem, Sumatra, with wet rice fields in the foreground. Orang-utans live in this forest.

Opposite Hill dipterocarp forest, Maliau Basin Conservation Area, Sabah. Wandering male orang-utans occasionally pass through this forest, but there is no breeding population.

Hill dipterocarp forest is found in the hill ranges that characterize much of inland Borneo between about 300 and 750 metres above sea level. In the absence of any human presence, extensive regions of this forest type without contiguous lowland forests in valleys do not seem to sustain orang-utan populations. Where there is also some adjacent lowland dipterocarp forest (for example, in Lanjak Entimau Wildlife Sanctuary, Sarawak), breeding populations of orang-utans may occur. Between about 750 and 1,300 metres above sea level on steep terrain is highland dipterocarp forest. Extensive regions of this forest type do not normally appear to sustain orang-utan populations, even where there have never been human populations. There is a unique situation in Sabah, however, where very small and highly endangered remnant populations of orang-utans exist in the mountainous Kinabalu Park and Crocker Range Park. Their activity appears to be concentrated on the border between this forest type and lower montane forest, with periodic visits down to hill dipterocarp forests. For the time being, it is assumed that these populations belong to the morio sub-species, which occurs more abundantly in the lowlands and swamps of eastern Sabah. These two mountain orang-utan populations must have been separated from the lowland populations of eastern Sabah for at least hundreds of years.

In Sumatra, the dipterocarp forest within the existing range of the orang-utan is usually regarded as a single type that changes gradually with increasing altitude above sea level. In general, as one goes higher into the hills, the overall forest structure becomes lower and tree species with fleshy fruits favoured by orang-utans become scarcer. At Ketambe, in the Alas Valley of the Leuser ecosystem on Sumatra, the altitude is 250–400 metres above sea level but the soils are mainly fertile alluvium with some slopes, and the forest continues uninterrupted to above 1,500 metres. Pioneering studies in the 1970s showed a population density of five to six orang-utans per square kilometre at Ketambe. Later studies show population densities ten times lower in the highest altitude dipterocarp forest. At the lowest altitude, where orang-utans exist naturally in dipterocarp forest in Sumatra, 30–100 metres above sea level at Sekundur on the east side of the Leuser ecosystem, orang-utans in primary and old, lightly logged forest are reported to exist at population densities of only one individual per three to ten square kilometres. The likely reason for such low numbers is not known, but may be a result of a long history of hunting in the lowlands. One study in the Leuser ecosystem suggested that, with no hunting pressure, higher orang-utan density is related to an abundance of large strangling figs and to less acidic soils. In general, Sumatran dipterocarp forests seem to support higher densities of large mammals than Bornean forests.

OTHER PRIMARY FOREST TYPES

Where the land form is flat and soils very sandy, the natural vegetation is usually heath forest, characterized by small trees and other plants with small thick leaves, of rather limited species diversity, and an overall low tree canopy. Under some heath forests are soils of extreme infertility, consisting simply of coarse, white quartz sand with a thin layer of dead leaves on top. A hard, iron-rich layer tends to develop in these soils at a depth of about one metre under the surface. This limits the vertical movement of water, so that heath soils tend to become very wet during rainy periods and very dry during rain-free periods. Heath forests can be found at almost any altitude above sea level, ranging from coasts up to more than 1,000 metres. On pale yellow sandy soils one also finds forests intermediate between heath and dipterocarp, with just a few dipterocarp species of low stature. Orang-utans do not seem to live entirely within any of these forests, although they may enter them from adjacent peat swamp

forest (for example, Central Kalimantan) or dipterocarp forest (for example in Sepilok Forest Reserve, Sabah). In regions where annual rainfall is rather high, and the land forms a low basin, sandy heaths tend to be waterlogged for much of the time, and a thin layer of peat builds up on top of the sand. In these areas, the vegetation is known as kerapah forest. A few relict patches of this forest type exist east of the Barito River south of Buntok, supporting a tiny, almost extinct population of orang-utans, which has never been investigated scientifically.

Two kinds of geological formation within the range of Bornean orang-utans produce soils that are so different from

Above *A rocky clearwater stream typical of the hill ranges and mountains of Borneo.*

Opposite *Primary forest on ultramafic soils, Tawai Protection Forest Reserve, Sabah. Orang-utans occasionally enter this forest from adjacent lowlands. An oil palm plantation is in the foreground.*

others that they have their own distinctive forest types. Limestone vegetation occurs on the numerous outcrops that exist in many parts of Borneo, with plant species composition and forest structure varying with the slope and form of the limestone and the amount of soil that has formed over the rock. Generally, limestone forest consists of just some of the tree species of dipterocarp forest, but with few dipterocarps. Surveys in the limestone karst forests of Sangkulirang Peninsula, East Kalimantan, indicated a population density of about one orang-utan per square kilometre, while adjacent extreme lowland dipterocarp forest had orang-utan densities five times higher, the highest ever recorded for that habitat type.

In Sabah, there are distinct forests on ultramafic or ultrabasic hills and mountains, where dark chocolate-coloured soil is derived from rocks rich in iron and other heavy metals. Orang-utans rarely range into these areas from adjacent dipterocarp forests.

On mountains in Sumatra and Borneo, above the dipterocarp forests, can be found lower montane forest, which has no, or very few, dipterocarps and instead is dominated by much smaller trees of the oak, laurel and myrtle families. In Sumatra, and on smaller mountains in Borneo, this forest starts at about 1,200 metres above sea level, but on larger Bornean mountains the start is higher. Some orang-utans include this forest within their range, but only at four known sites: in the Leuser ecosystem in Sumatra; on Mount Kinabalu and in the Crocker Range, both in Sabah; and in the Gunung Palung National Park in West Kalimantan. The latter site is unique within the distribution of wild orang-utans, having peat swamp, dipterocarp and montane forests in close proximity.

There are numerous oak species in Borneo, and their habit of producing large quantities of fruits outside the usual forest fruiting peaks may be one reason why orang-utans can survive in these zones. In some mountain ranges in Borneo where orang-utans are absent, the zone between highland dipterocarp and lower montane forests contains an additional vegetation type, Agathis forest, dominated by large coniferous trees of the genus *Agathis*. At altitudes higher than 2,000 metres above sea level is upper montane forest, consisting of stunted bushy trees, not used by orang-utans

Right *Montane forest, Sabah. Prominent features include numerous small trees, few lianas, abundant mosses, and a thick carpet of slowly decaying leaves.*

DISTURBED FORESTS

All the forest types described so far are natural habitats that develop over thousands of years in the absence of clearance or major disturbance by humans. Until a few decades ago, the majority of Borneo and Sumatra was covered in these natural forests. Human communities, centred mainly on the coasts and along valleys, cleared forests locally for cultivation of non-irrigated rice, but used old growth forests only for harvesting of forest products such as meat, palm fronds or natural products such as resins, for trade. That has changed drastically since the mid-20th century. Now, extensive areas have been converted to plantation, notably of oil palm, rubber and fast-growing trees for wood production. Orang-utans cannot live in any kind of plantation because there is simply not enough food for them

year round. They feed on the growing shoots of newly-planted oil palms, but do not enter mature plantations to feed on the fruits. The most extensive habitats in Borneo and parts of Sumatra now, however, are not plantations but are what are generally termed logged forests, meaning forests where a large proportion of the commercially valuable timber trees have been removed. Logged forests now represent a wide spectrum of types, based partly on their original characteristics before disturbance, and then on the degree of disturbance incurred over the past few decades.

A continuous spectrum of logged forest now exists, ranging from sites where just a few big trees were removed years ago, to sites where every tree that can be sold has been removed by people who are not concerned about forest regeneration. In order to describe logged forests, we need to understand that

the original primary or old growth forests described above contain a wide variety of tree species that are ecologically adapted to conditions of shade, without exposure to full sunlight, but with the constant high humidity that exists under full forest cover. Most of the tallest dipterocarp and legume trees that form the upper canopy of undisturbed tropical rainforest started their life as seeds on the ground under conditions of low or uncertain light intensity and high humidity. These tall trees in turn provide conditions of shade and humidity to the numerous other tree species that remain small, even when old and mature.

Until around the 1970s, the types of tall forests that were exploited for timber production (mainly dipterocarp and some peat swamp forests) were managed under a simple system recommended by foresters. The basic rules were to fell and remove trees of good condition above a minimum trunk

Opposite *View of a typical Bornean primary dipterocarp forest, with a dense but uneven canopy of tall trees, including some old standing dead trees.*

Above *View of a typical Bornean logged dipterocarp forest, with many of the tall trees removed, leaving non-commercial species, and new space for seedlings and young trees of commercial species to grow.*

diameter at breast height, usually either 50 or 60 centimetres. The act of logging was what foresters called silviculture, a convenient means of obtaining timber while enhancing conditions to replant trees, because there are already ample natural seedlings, saplings and immature trees, suddenly provided with the conditions they need – extra space and light – to boost their growth. The only catch is that the people who want to produce timber have to wait a very long time for the next round of tree cutting. According to the original forest composition and soil type, this might vary between 35 and 80 years. During the 1980s, probably the most extensive habitat in much of Borneo and Sumatra was logged dipterocarp forest, which had been logged just once. If left alone, and not affected by fires during El Niño droughts, it would have regenerated to tall dipterocarp forest.

But the inevitable happened. Many people had long realized that logging natural forests for timber was a money-spinner. No one had invested in the original, natural forests, so the timber was essentially 'free'. All that was needed was human labour, chainsaws and heavy machinery to cut and remove the logs to buyers. A whole series of factors came into play that led initially to perfectly legal, but premature, second and third logging of logged forests, and then to taking of trees less than 50 centimetres diameter. In some

cases, the authorities reduced the minimum allowable cutting size, either because there were plans to convert the forests to plantations, or to ease shortfalls in tax revenue. In many cases, logging contractors cut undersized trees, claiming that they were the narrow top end of bigger trees, or cut small trees outside their licence areas and put the blame on others. In many areas, men without any alternative opportunities to earn regular income cut, and still cut, any convenient trees that could be sold, month after month, year after year. Sooner or later there were no more big trees, so they cut the smaller ones. In other words, a whole gamut of factors have acted together to undermine the original system of sustainable forest management. The factors include low investment costs, quick returns, too many contractors, revenue needs in a developing economy, rural poverty, weak governance in some areas, narrow-based economies that have few or no options to absorb a growing workforce, and a growing local and global demand for wood.

In Borneo and Sumatra, long (El Niño) droughts mean virtually no rain for a continuous period of between four and nine months in forest ecosystems that are adapted to coping with dry periods of normally no more than two or three months. El Niño droughts are not a new phenomenon. A combination of radiocarbon dated charcoal in forest soils, tree

rings in Java teak trees since the 16th century, and colonial records for Kalimantan, Sarawak and Sabah dating from 1747 indicate that long droughts have occurred unpredictably at many times in the region over the past few hundred years. Some scientists have speculated that the differences in tree species composition seen in primary forests of Borneo reflect the aftermath of these occasional catastrophic droughts, rather than the 'normal' conditions of rainfall that characterize most years. We might equally speculate that such droughts may have wiped out orang-utans from large areas, thereby accounting for some of the big gaps in their distribution, especially in regions far from large rivers and swamps. Rainfall records show that droughts in Borneo in 1877–78, 1902–03 and 1914–15 were at least as severe as those of 1982–83 and 1997–98. Contemporary accounts mention forest fires and a pall of smoke during the early long droughts. But the 1982–83 and 1997–98 droughts were different and far more damaging, mainly because of the combination of vast amounts of inflammable dead wood in logged forest and the ubiquity of human presence. It is estimated that over 4.3 million hectares of logged forest burned in Borneo during 1982–83, and another 3.5 million hectares (including some primary hill dipterocarp forest, but excluding areas already degraded from 1982–83) in 1997–98.

The areas that suffered most from these fires were non-swampy and peaty lowlands, the areas that had formerly been covered with lowland dipterocarp forests and peat swamp forests, and which had been the prime habitats for Bornean orang-utans.

The widespread logging over a few short decades, coupled with periodic drought and fire damage, has resulted in a logged forest that no longer represents a uniform habitat type. Where once there was simply dipterocarp forest, there is now a patchwork of different forest conditions.

In order to describe more clearly the situation and its implications for orang-utans, as well as for biodiversity and for future timber production, we can imagine two extreme types of

Opposite *Kumai River and Kumai town, Central Kalimantan, set in the vast peat swamps that characterize much of southern Borneo. The forest here has been subject to droughts and fires in 1982–83 and 1997–98.*

Above *Peat swamp forest, Central Kalimantan, which has been subjected to recent (2006) localized fire.*

FOREST CLEARANCE, FIRE AND GREENHOUSE GAS EMISSIONS FROM PEAT SWAMPS

Up to about the 1950s, hardly anyone ventured into peat swamp forests because the waterlogged ground could not be cultivated, and even walking on it was fraught with problems, as one stubbed feet on sharp stilt roots or sank into metre-deep potholes of wet leaves. In the 1950s it was found that the peat swamps were full of medium-sized trees of great commercial value that could be cut and removed conveniently via small canals dug though the forest to a river. As these forests lost their obvious economic value once the main commercial trees had been cut, there emerged a whole series of threats to the peat swamp ecosystems. The former president Suharto of Indonesia was persuaded to clear one million hectares of peat swamp forest in south-eastern Central Kalimantan in order to develop a new giant wet rice-farming region. Much of the forest was cleared, drainage canals dug, and some attempts made at cultivation. As scientists had predicted, it was found that rice cannot be grown economically in these swamps and the canals effectively dried the peat to convert it to a massive tinder box. A public highway was built across this massive swamp, linking the provincial capital of Palangkaraya to other towns, but making the entire swamp all the more susceptible to fires. Similar trends occurred in many other peat swamp areas in Kalimantan and Sumatra. In some cases, small-scale farmers clear and burn the fringes in an attempt to eke out a living. Several of the larger Indonesian and Malaysian oil palm companies have banked their future growth on expanding large-scale plantations onto extensive peat soils in Kalimantan, Sumatra and Sarawak. Long experience on smaller tracts of peat soils in Peninsular Malaysia has shown that oil palm can indeed be grown successfully with good yields. The much larger peat areas of Borneo and Sumatra present greater challenges, including how to retain optimum water levels at all times, how to keep out fire during dry periods, how to prevent mature palms falling over in the soft peat, how to get sufficient plant micro-nutrients into soils with no mineral base, and whether the higher management costs of peat-based plantations will prove financially attractive in the longer term.

The burning and subsequent smouldering of exposed peat where there was formerly peat swamp forest is by far the biggest source of the thick haze that envelops parts of Indonesia, Malaysia, Brunei and Singapore during drier years. The haze causes sickness, potentially long-term damage to lungs, closed schools, cancelled flights, and massive losses to local economies.

There are also longer-term concerns. One study, based on short-term data, estimated that the seven million hectares of peat swamp in Kalimantan emit less than 0.3 per cent of global greenhouse gas emissions. But once exposed by heavy logging, drainage, fire or clearance, peat soils oxidize and start to release large volumes of carbon dioxide even if there is no further fire. Some calculations suggest that, if the peat lands of Indonesia are not managed with care, carbon dioxide emissions will eventually cause Indonesia to become the third global emitter of this greenhouse gas, after the USA and China.

logged forest. In heavily logged forest or secondary forest few, if any, original large trees remain. Those that do remain are of very poor form, or non-commercial species, and the bulk of the immature dipterocarps and their seedlings have died as a result of disturbance and exposure. Seventy per cent or more of the tree cover now consists of pioneer tree species, those species that are ecologically adapted to open conditions with regular strong sunlight. Common pioneer trees are members of the genus *Macaranga*, *Anthocephalus chinensis* (also known as *Neolamarckia cadamba*), *Trema* species, *Pterospermum* species, *Octomeles sumatrana* and *Duabanga moluccana*. In the other

sort, which can be called logged dipterocarp forest patches, the main elements of the original forest structure remain in place. There are still some large dipterocarp trees to act as seed sources, along with immature dipterocarps and a diversity of the

Opposite (top) *In the absence of alternative sustainable livelihoods, people have tried to settle and farm in the vast peat swamps of Central Kalimantan, made accessible in recent decades by drainage canals, logging tracks, fire and roads.*

Opposite (bottom) *In recent years, attempts have begun to restore the damaged Central Kalimantan peat swamp ecosystem by regulating water flow in 4,000km of drainage canals dug between 1996 and 1998.*

Left *Fruiting dipterocarp tree (Shorea species). Individual dipterocarp trees tend to fruit once every few years, in unison, but some ecologists have reported a trend since the 1990s of fewer trees producing fewer fruits and less successful germination of the seeds.*

Opposite (top) *Recent logging of lowland dipterocarp forest that was first logged in the 1970s.*

Opposite (bottom) *Regenerating logged forest, lower Kinabatangan, Sabah.*

original middle-canopy non-dipterocarp trees. This forest may cover thousands of hectares in places, for example in some protected areas established after logging, or where there are research projects, or on extensive steep slopes where it is difficult to fell and remove trees. Or, more commonly, they may be simply islands of just a few to a few tens of hectares surrounded by secondary forest. If left alone, these patches will not only regenerate as dipterocarp forest but will act as natural seed sources for recolonization and regeneration of the surrounding heavily logged areas.

Initial observations made in the 1980s, mainly based on counting orang-utan nests (see page 42, Surveying orang-utan populations by counting their nests), suggested that orang-utans tend to stay in patches of undisturbed forest within a landscape of logged dipterocarp forest, but that the overall population density of orang-utans remained remarkably stable after one round of logging. This is perhaps not too surprising, because the

tall dipterocarp trees which represented the bulk of commercial trees cut by loggers are not important as orang-utan food sources. Dipterocarps produce crops of hard seeds only rarely, most often when there are numerous other trees producing preferred fruits. Where forest has been logged more than once and where fires have occurred, a similar pattern emerges, with orang-utans seeming to concentrate in the best remaining patches of forest.

This appears to imply that orang-utans are safe, and that they can live happily in heavily logged forests. In fact, we are a long way from being so confident. Heavily logged forests are still tropical rainforest, but they are damaged and very unlike the forests which previously sustained orang-utans over many generations. The more a forest is logged, the more orang-utan food trees disappear, not primarily because they are removed for timber, but because they are knocked over when the timber trees fall, and as bulldozers pull out logs from the forest.

SURVEYING ORANG-UTAN POPULATIONS BY COUNTING THEIR NESTS

The earliest researchers of orang-utans in the tropical rainforest realized that much information can be obtained from counting the nests of orang-utans.

In a state-wide survey of the status of orang-utans conducted by Sabah Forestry Department with WWF-Malaysia in the mid-1980s, the idea was developed of using helicopters as a means of counting orang-utan nests over wide areas. The initial idea came from two members of the Sarawak Forestry Department, who in 1981 flew over Lanjak-Entimau Wildlife Sanctuary and realized that they could easily spot orang-utan nests in the crowns of trees in the forest below. From 1985 to 1988, much of Sabah was surveyed with the help of the Royal Malaysian Air Force and the government-owned Sabah Air. Since every adult orang-utan normally makes one nest per day, it was assumed that the number of orang-utan nests seen per kilometre of forest overflown was a measure of the relative numbers of orang-utans present. Amazingly, it was found that the highest densities of nests were in heavily logged extreme lowland dipterocarp forest and in freshwater swamp forests. Furthermore, it seemed that the density of nests was linked more to the altitude of the land than to the condition of the forest. Putting together all the results, it was found that, in general, the lower the land the more nests were seen. However, there was concern that the number of nests seen from the helicopter was also dependent on the condition of the forest. The more disturbed the forest, the more open is the tree cover, and the easier it is to spot nests. Between 2001 and 2003, a similar Sabah-wide survey, consisting of extensive ground and helicopter counts of nests, was conducted by the Sabah Wildlife Department with the Kinabatangan Orang-utan Conservation Project supported by the French NGO, Hutan. Those surveys provided a more refined picture, showing that proportionately more orang-utan nests can be seen in the more damaged forests, and also that there are odd differences from place to place, but that overall the highest numbers of orang-utans are indeed in the lowland and swamp forests.

Most orang-utan field researchers continue to grapple with deciding the best ways to get the most, and most reliable, data from counting nests. Kinabatangan Orang-utan Conservation Project in Sabah has shown that the average orang-utan makes only one nest per day in the lower Kinabatangan (fewer than in other study sites, where the average is greater than one, because orang-utans sometimes make daytime siesta nests), that nests decay much more quickly in some tree species than in others (especially in the fast-growing species which characterize heavily damaged forests), and that nests tend to be made in small trees that may not be visible from a helicopter. On the ground, researchers find that it is much easier to spot nests from the side. They often cannot see nests directly above, over a trail. That means that the straight-line trails preferred by theoretical biologists may not be the best choice in real-life forest conditions.

As we shall see in other chapters, orang-utans are like people in many aspects of their biology. If anything, their children are even more demanding than human children. Any form of prolonged stress can affect their health. Orang-utans that travel on the ground – as they have to do in extensively damaged habitat – tend to have a wider array of debilitating diseases than those in primary forest. Eating poor quality food for many years may sap their energy, health and reproductive success. In other words, orang-utans in extensive, heavily logged forest can be compared to subsistence farmers in a marginal landscape, relying on a monotonous diet that may be poor in minerals and vitamins, plunged into dire circumstances if the staple crops fail. In primary forest, orang-utans have daily access to a food 'supermarket', with some degree of choice even in bad times, and the opportunity to relieve stress by resting in shade and coolness.

Opposite (top) *Late every afternoon, mature and older adolescent orang-utans build a simple nest in which to sleep at night. Normally, a nest is abandoned after one night's use, decaying over a period of many months.*

Opposite (bottom) *Nests are built by bending and snapping small branches and twigs to form a platform.*

Above *Orang-utan nests are readily visible from a helicopter. The nest below centre of this picture is quite fresh, while the one above right is old and decaying.*

Right *This orang-utan has made a nest in a fruiting tree, convenient for tomorrow's breakfast.*

Opposite *Young orang-utans may play at snapping off twigs and building play nests, but they sleep with their mother in a single nest up to six years of age.*

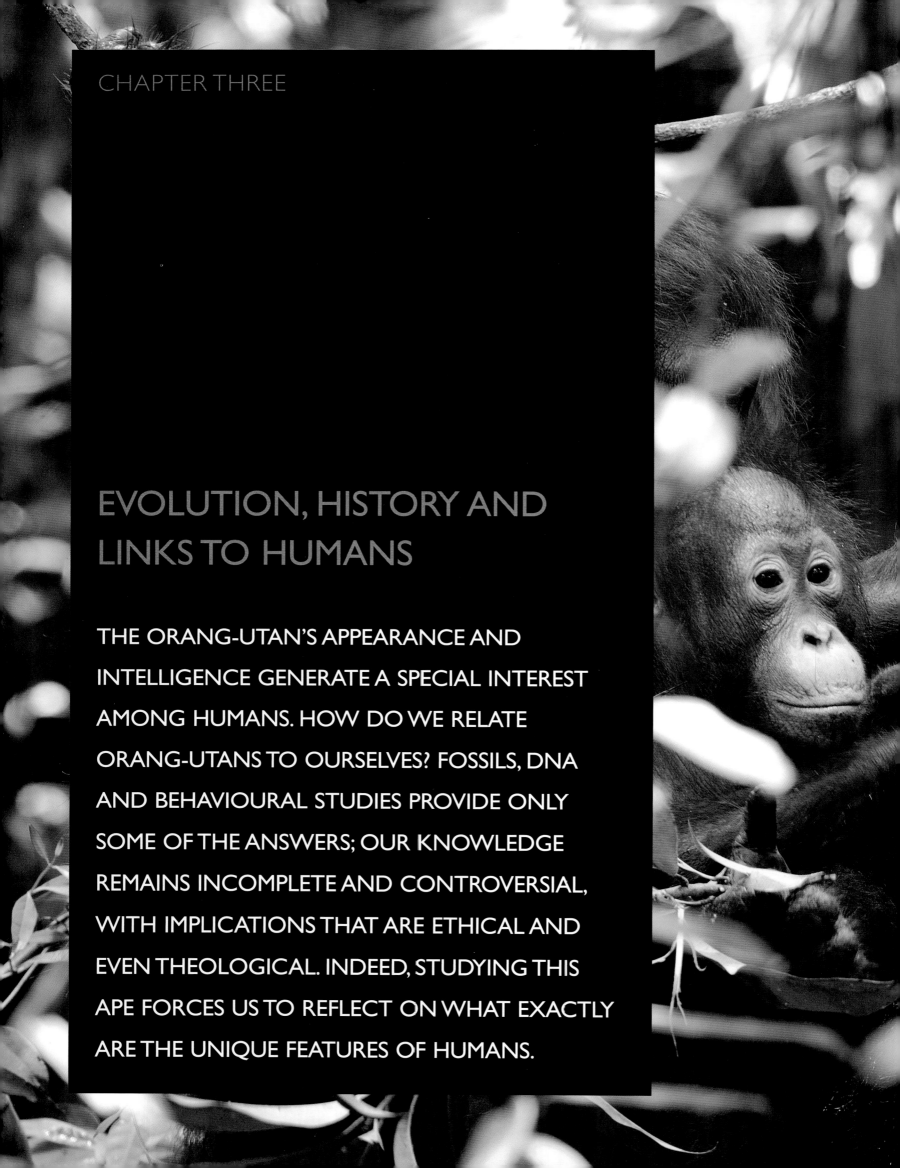

CHAPTER THREE

EVOLUTION, HISTORY AND LINKS TO HUMANS

THE ORANG-UTAN'S APPEARANCE AND INTELLIGENCE GENERATE A SPECIAL INTEREST AMONG HUMANS. HOW DO WE RELATE ORANG-UTANS TO OURSELVES? FOSSILS, DNA AND BEHAVIOURAL STUDIES PROVIDE ONLY SOME OF THE ANSWERS; OUR KNOWLEDGE REMAINS INCOMPLETE AND CONTROVERSIAL, WITH IMPLICATIONS THAT ARE ETHICAL AND EVEN THEOLOGICAL. INDEED, STUDYING THIS APE FORCES US TO REFLECT ON WHAT EXACTLY ARE THE UNIQUE FEATURES OF HUMANS.

THE ORANG-UTAN'S PLACE IN EVOLUTION

This book takes the view that evolution of living organisms has occurred by a process of natural selection that commenced sometime in excess of three billion years ago. The evidence is overwhelming from all aspects of the natural world, especially the existence and periodic finding of new fossils, coupled with the behaviour of DNA (deoxyribonucleic acid, the biochemical that contains the genetic instructions for the development and function of living things). It is not impossible that divine intervention has occurred to modify evolution from time to time. Belief in such interventions is a matter of faith, as there is no scientific supporting evidence either for or against it. What is not tenable, based on all or any available evidence, is the idea that humans, or orang-utans, or any other organism, have been created spontaneously at some specific time in prehistory.

Whatever one's belief, the idea of the existence of an entity that we call species is almost self-evident. A species is a basic unit of what is nowadays called biodiversity. Of the many possible definitions of the term 'species', the following is one of the simplest: *all the individual organisms of a natural population which are able to interbreed*. In scientific classification, a species is assigned a two-part name – the genus, followed by a second word which defines the species. It is a scientific convention to write the genus and species names in italics. The Bornean orang-utan is thus named *Pongo pygmaeus*.

Orang-utans are classified zoologically, along with other apes, monkeys, prosimians and humans, within an Order called Primates, which has its origins in the Eocene epoch about 55 million years ago. Over recent decades, there have been significant changes of opinion among researchers on the way in which orang-utans, humans and other primates are believed to be related to one another. Our changing perceptions are based partly on new evidence from fossil discoveries and anatomical comparisons, and more especially on comparison of serum proteins and DNA sequences.

The most general current view among zoologists is that orang-utans, gorillas, chimpanzees, bonobos and humans fall within a single zoological family, the great apes, known scientifically as the Hominidae. Originally, the name Hominidae referred to the living human species and extinct human-like relatives, while the other great apes were placed in a separate family, the Pongidae. Now, with a whole array of available evidence from fossils and DNA, it seems that all the living great apes and humans, along with a variety of extinct great apes, including hominoids of the genus *Homo*, evolved from a single, common ancestral ape. From a scientific angle, it makes sense to place all these species, together with their long-extinct common ancestor, into one family – Hominidae.

In classical zoology, as well as modern classifications based mainly on DNA, animal families are sub-divided into sub-families, genera (singular genus), species and sub-species.

The great ape family breakdown, showing only species that are still living, would appear like this:

FAMILY HOMINIDAE (Great Apes)	
SUBFAMILY PONGINAE Genus *Pongo*	
Species	Bornean Orang-utan *Pongo pygmaeus* Sumatran Orang-utan *Pongo abelii*
SUBFAMILY HOMININAE Genus *Gorilla*	
Species	Western Gorilla *Gorilla gorilla* Eastern Gorilla *Gorilla beringei*
Genus *Pan*	
Species	Common Chimpanzee *Pan troglodytes* Bonobo *Pan paniscus*
Genus *Homo*	
Species	Human *Homo sapiens*

are more closely related to each other than are chimpanzees to orang-utans. The similarity of DNA sequences represents an element of time – how long ago the different species separated from a common ancestor.

While DNA studied in laboratories has revolutionized many aspects of science in recent decades, fossil discoveries continue to help us understand the origin of orang-utans and other apes.

In the Ponginae sub-family, some 12.5 to 8.5 million years ago, was *Sivapithecus*, a genus of ape with facial and palate features that show similarities with orang-utans, living in what is now India and Pakistan. Also in this sub-family was the genus *Lufengpithecus*, living in what is now southern China and Thailand between eight and two million years ago, a period during which many types of hominoid became extinct elsewhere in Europe and Asia. Both those genera would have lived in conditions drier and cooler than we see in Borneo and Sumatra today. *Khoratpithecus piriyai*, which occurred in Thailand between seven and five million years ago, seems to be the closest of the extinct fossil Ponginae to modern orang-utans, perhaps not surprisingly as it would have lived in a tropical climate similar to that of modern Borneo and Sumatra. The aptly named *Gigantopithecus*, a three-metre tall creature in the same sub-family, which lived from five million years to about 100,000 years ago, occurred in China, India and Vietnam, probably including bamboo in its diet. Some people suggest that this animal represents the origin of stories about the yeti, a giant ape that some claim still lives in the Himalayas.

Fossil and DNA evidence suggests strongly that the centre of great ape evolution was Africa, with the genetic lines that led to orang-utans and gibbons having separated in the order of 20 million years ago. The genus *Homo*, in the form of *Homo erectus*, appears to have entered Asia from Africa about two million years ago.

In this book, the living orang-utans are regarded as belonging to two species, with three sub-species of the Bornean orang-utan, but some researchers hold alternative views.

It seems that the apes, defined in simple terms as primates having no tail and relatively large brains, split off as a new group from the Old World (African and Asian) monkeys about 25 million years ago. The ape line split at some time between 15 and 18 million years ago into lesser apes (gibbons) and great apes. After that, many forms of great apes evolved, most of which are now extinct, leaving the seven living species. DNA evidence suggests that the orang-utan (genus *Pongo*) split off at some time between 11 and 14 million years ago and the gorilla (genus *Gorilla*) 7 million years ago, while the chimpanzee-human (genera *Pan* and *Homo*) split started about six million years ago and continued via several extinct genera to the evolution of *Homo* about two million years ago. Looking at evolution in this way helps to show why scientists say that humans and chimpanzees

Left *All species have a two-part scientific name, normally written in italics or a typeface different from the surrounding text. The Bornean orang-utan is named* Pongo pygmaeus.

Opposite *The Sumatran orang-utan is named* Pongo abelii. *The genus Pongo is believed to have split about 11–14 million years ago from another line, which evolved to gorillas, chimpanzees, humans and other, extinct, apes.*

There is evidence that an enormous volcanic explosion, which occurred about 73,000 years ago, led to massive global weather and climate change, either exterminating or almost exterminating many species, including the great apes. The explosion occurred at Toba in northern Sumatra, near to the last modern surviving populations of Sumatran orang-utans. It has been suggested that *Homo sapiens* was exterminated throughout Asia as a result of that event, but that a small population survived in Africa, spreading back into Asia about 45,000 years ago.

In the 1930s, Gustav von Koenigswald, a German palaeontologist, travelled around Asia in search of fossil teeth and bones of mammals. In Chinese medicine shops in Peking, Canton, Hong Kong and Manila, in the Philippines, he found 'dragon's teeth', which on closer examination turned out to be the semi-fossilized teeth of mammals, including those of orang-utans. All the teeth found by von Koenigswald seem to have come originally from caves and rock fissures in what is now southern China and North Vietnam. Later, more orang-utan teeth were discovered at various sites on mainland South-east Asia, mostly in cave sediments, and some mixed together with the teeth of prehistoric giant pandas, mountain goats and other species that do not occur in tropical rainforests. These teeth are probably all several hundreds of thousands of years old.

Many of these ancient orang-utan teeth are nearly 40 per cent larger than those of modern orang-utans, and one might therefore suppose that early orang-utans were proportionately larger. Orang-utan teeth of similar age but of varying sizes, some large and some like those of modern orang-utans, have been found on the island of Java in Indonesia. There are also prehistoric orang-utan teeth from caves in Borneo and Sumatra, some larger than those of modern orang-utans, but none are more than 40,000 years old and most are more recent. It seems clear that, half a million years ago and perhaps earlier, orang-utans were widespread throughout much of South-east Asia, and that they later became extinct in most regions. The distribution and sizes of the ancient orang-utan teeth suggest that the original orang-utans were large, and that smaller ones evolved later as an adaptation to life in humid tropical forests. Alternatively, there may have been two kinds of orang-utan from very early times, a large form in the mountains and a smaller one in the lowlands.

These changes in orang-utans happened largely independently of human populations. Humans in Asia 40,000 years ago were 'pre-Austronesian', while extant Bornean and Sumatran people are of Austronesian stock, believed to have originated from Taiwan less than 10,000 years ago.

THE LIVING ORANG-UTAN SPECIES

Although many zoologists now regard the Bornean and Sumatran orang-utans as separate species, it is really a matter of opinion as to whether the two forms are best regarded as full species or sub-species of a single species. In fact, for most of the 20th century, it was normal for the orang-utan to be classified as a single species, with two sub-species: Bornean (*Pongo pygmaeus pygmaeus*) and Sumatran (*P. p. abelii*).

Strong evidence favouring the idea that the Bornean and Sumatran orang-utans are best regarded as different species, *Pongo pygmaeus* and *Pongo abelii* respectively, came initially from Xiufeng Xu and Ulfur Arnason of the University of Lund, who in 1996 published a paper which indicated that the degree of molecular difference between the Bornean and Sumatran orang-utan is greater than that between the common and pygmy chimpanzees (which are now regarded as different species), and similar to that between the horse and the donkey, and the fin and blue whales (which have always been regarded as different species).

Managers and keepers of zoo orang-utans point out that there are records of hybrid orang-utans, the offspring of Bornean and Sumatran orang-utans mating and producing second generation hybrid offspring. In classical zoology, this could be viewed as indicating that the two forms of orang-utan are best considered as a single species, which in their wild state just happen to be separated now by the 150-kilometre wide Karimata Strait between Borneo and Sumatra.

DNA studies by various researchers suggest that the Bornean and Sumatran orang-utans may have been separate genetic units for a long period, with estimates ranging from less than one million to as long as five million years. One can reasonably argue that populations that have been separated for such a long period deserve to be regarded as separate species, even though they can interbreed. But the islands off South-east Asia are part of an extensive continental shelf, called Sundaland, which has been joined intermittently to mainland Asia for long periods, and as recently as 12,000 years ago. It would be surprising if the two populations had remained totally separate for millions of years.

Opposite *Based on fossils found in caves in Borneo and Sumatra, living orang-utans appear to be smaller than those of the past.*

Above *Bornean and Sumatran orang-utans have hybridized in zoos; this mature male has Sumatran and Bornean features.*

In a study of the evolutionary history of modern primates of Sundaland, Terry Harrison and Jessica Manser of New York University and John Krigbaum of University of Florida provide a story suggesting that the orang-utan spread from its mainland Asia origins through Sundaland about 2.8 million years ago, when sea levels were at least 100 metres lower than they are now. As sea levels rose and declined, and suitable habitats receded, starting about 1.8 million years ago, the widespread population fragmented and became isolated.

Other DNA studies reveal other surprises. One is that the Sumatran orang-utan, although now confined to a very small area, and physically distinguishable from Bornean orang-utans, has much greater genetic diversity within the species than does the Bornean orang-utan. More surprisingly, both species of orang-utan have a greater within-species genetic diversity than do humans. What are we to make of that – other than support for the idea that humans nearly became extinct about 70,000 years ago after the Toba explosion, and have evolved from that genetic bottle-neck over a geologically short time-span into the great physical variety that we see today?

One possibility is that the orang-utans living nearest to Toba at the time of its explosion were wiped out, and that small outlying populations with genetic differences survived in Borneo, Java and mainland Asia. Orang-utans from these three sources may have spread back later into what is now Sumatra, interbreeding and contributing to the genetic diversity.

Orang-utans were apparently still widespread throughout South-east Asia during the Pleistocene epoch (about 1.8 million to 10,000 years ago, a period of repeated glaciations during which sea levels rose and fell, and the climate changed in tandem). Partial fossils of teeth and bones of *Pongo* have now been found in many sites in southern China, Vietnam, Laos, Cambodia, Thailand and Java, as well as Borneo and Sumatra. In some of those fossils, the teeth are different in form from those of modern orang-utans, and have been described as coming

Left (top) *Although available literature suggests that there are no wild orang-utans between the Barito and Mahakam Rivers, a very small and nearly extinct population does exist there, possibly representing an undescribed fourth Borneo sub-species.*

Left (bottom) *Pongo pygmaeus pygmaeus (mature female) – the sub-species in Sarawak and West Kalimantan, north of the Kapuas River.*

Opposite *Pongo pygmaeus morio (mature female with infant) – the sub-species in Sabah and East Kalimantan.*

Above *Sabah and East Kalimantan orang-utans (the former in this picture) are normally regarded as belonging to a single sub-species, but populations in the two regions may have been geographically separate for many thousands of years.*

Opposite *For gibbons and leaf monkeys, differences in the loud calls of males help to distinguish between species and sub-species, but the loud, long calls of male orang-utans vary greatly between individuals in all wild populations.*

from a different species, *Pongo hooijeri* (named after D.A. Hooijer, a prolific Dutch palaeontologist of the mid-20th century). In others, the teeth and bones have the same shape as those of modern orang-utans, and indeed are in some cases intermediate between those of modern Bornean and Sumatran orang-utans. Thus, the fossil and sub-fossil orang-utans are regarded as being close to living orang-utans, and are named as different sub-species of *Pongo pygmaeus*. But even though the body evolved to smaller size, the skulls of fossil orang-utans tend to be relatively large, and the incisor, premolar and molar teeth are very large. The only complete skeleton of an adult prehistoric orang-utan, identified as a form of *Pongo pygmaeus*, was found in 1997, in an underground cave in northern Vietnam. Its age is unknown but presumed to be late Pleistocene (that is, between 125,000 and 10,000 years old).

Clues as to the origins and relationships of the living wild orang-utan populations normally come from fossils and DNA studies. Using a different approach, however, Perth Zoo exotic mammal curator, Leif Cocks, compared aspects of breeding among 1,678 orang-utans held or born in zoos globally over the period 1946–94. Hybrids had significantly lower survival rates and higher infant mortality than did either the pure Bornean or Sumatran orang-utans. Also, hybrid infants experienced significantly higher frequency of rejection by their mothers.

Since there was no evidence that the worst zoos had the most hybrids, or that zoo staff were taking less care of hybrids, he suggested that the most likely explanation for these findings was that hybrids are weaker. This would support the idea that Bornean and Sumatran orang-utans belong to significantly different populations.

But Cocks also found that the Bornean orang-utans had much higher rates of infertility than did either the Sumatran or hybrid orang-utans. He suggested that this may have been the result of mixing individuals from incompatible populations from different parts of Borneo. The implication may be that orang-utans from different parts of Borneo are genetically distinct, in some ways as distinct as the Bornean and Sumatran forms.

Returning to more classical evidence from anatomy and from DNA comparisons, the generally accepted view is that there is a single form of the Sumatran orang-utan, with no division into sub-species, and three distinct sub-species of Bornean orang-utans, as follows:

- The southern sub-species, *Pongo pygmaeus wurmbii*, in Central Kalimantan and West Kalimantan (south of the Kapuas River), the largest of the three.

- The north-eastern sub-species, *Pongo pygmaeus morio*, in Sabah and East Kalimantan, the smallest of the three.

- The north-western sub-species, *Pongo pygmaeus pygmaeus*, in Sarawak and West Kalimantan (north of the Kapuas River), intermediate in size.

Based on DNA results, some researchers suggest that the orang-utans in Sabah and in East Kalimantan may be distinct.

Marina Ross of Hannover University took a different approach to comparing the various forms of orang-utan, by analyzing long calls of mature males. Her results did not conflict with the above classification, but she found great variation in the details of these calls, both within and between the single Sumatran and three Bornean groups.

Although all available published literature suggests that there are no wild orang-utans living between the Barito River and Mahakam River (the south-east part of Borneo, covering South

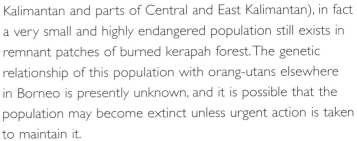

Kalimantan and parts of Central and East Kalimantan), in fact a very small and highly endangered population still exists in remnant patches of burned kerapah forest. The genetic relationship of this population with orang-utans elsewhere in Borneo is presently unknown, and it is possible that the population may become extinct unless urgent action is taken to maintain it.

Cam Muir, of the University of Hawaii, and colleagues take a rather different view, regarding all orang-utans as belonging to a single species. They point to the high genetic diversity within the Sumatran form, as great as that between the Bornean sub-species. They also suggest that, while the three Bornean sub-species appear to share a single common ancestor, the current Sumatran orang-utan population is derived from at least two, formerly distinct and separate, populations which have subsequently merged. Support for that idea comes from the observations of Herman Rijksen, who conducted the first detailed study of wild Sumatran orang-utans at Ketambe between 1971 and 1975. He describes the two forms as 'dark-haired, long-fingered' and 'light-haired, short-fingered' (see box on page 59: Differences between and within Sumatran and

Bornean orang-utans). Given the limited geographical distribution of the Sumatran orang-utan, and the fact that two (or more) ancient lineages have interbred over thousands of years, zoological conventions do not allow us to differentiate Sumatran orang-utans into two sub-species.

Above (left) *Mature Sumatran male.*

Above (right) *Mature Bornean male orang-utan, feeding on mud, possibly to obtain toxin-absorbing kaolin clay (the scars on the cheek flanges are signs of a serious fight with another male).*

Opposite (left) *Mature Bornean female.*

Opposite (right) *Mature Sumatran female.*

DIFFERENCES BETWEEN AND WITHIN SUMATRAN AND BORNEAN ORANG-UTANS

Orang-utans from Borneo and Sumatra are very similar, but differ in some physical and behavioural details. Bornean orang-utans are usually dark orange-coloured when young, becoming a dark rusty-red or chocolate with age. Some old ones are almost black. The body hairs of Bornean orang-utans are stiff and shiny, while those of the Sumatran orang-utans are softer and duller. Mature male Bornean orang-utans have a black, square-shaped face, with broad, protruding cheek flanges, deeply puckered forehead and a short beard. The face of the Sumatran male is longer, greyish, with less prominent cheek flanges, and a longer beard, usually complete with a moustache. Observations of orang-utans both in the wild and in zoos indicate that Sumatran orang-utans are generally more sociable among themselves than are their Bornean relatives.

Complicating the differences between Bornean and Sumatran orang-utans, Herman Rijksen describes two forms of the Sumatran orang-utan. One has relatively dark hair, dark skin, slender limbs, fingers and toes, with a well-developed thumb and hallux (big toe) with nails. The character is described as introverted, unpredictable, prone to anger, highly intelligent, and interested in solving technical problems. The other form is more heavily built, with paler hair, grey-brown skin, short, thick fingers and toes, stumpy thumbs and the hallux usually without nails. Its character is described as sociable, placid, extroverted, friendly, prone to fright, and playful when confronted with problems. During his observations of over 80 Sumatran orang-utans at Ketambe, Rijksen noted that the latter form is more common but that there are also intermediate forms between these two extremes. Both forms, especially the latter, normally have whitish hairs on the face and groin, a feature missing in the Bornean orang-utans. Zoo hybrids of mixed Bornean and Sumatran descent are intermediate in appearance, but usually have whitish facial hair.

HUMAN PERCEPTIONS AND RELATIONS WITH ORANG-UTANS

Cave deposits in Vietnam show that a predecessor of humans, *Homo erectus*, lived alongside orang-utans about 500,000 years ago. Fossils at Niah Caves in Sarawak show that humans were present in Borneo by around 45,000 years ago. It is probable that these prehistoric people obtained carbohydrate foods from forest roots, tubers and stems. But tropical rainforest plants offer very little to humans, with their strict dietary requirements, in the way of edible parts. The main human food of recent millennia in Borneo is rice, but this crop grows in the open. Cutting down tropical rainforest to grow rice requires axes, which suggests that rice is a relatively recent crop in the history of human settlement. It is likely that, before the last few thousands of years, most people in the forests of South-east Asia lived mainly on flesh. Excavations in several caves in the region show that, in early times, people were fishing, collecting molluscs and hunting all manner of animal life, including orang-utans. At Niah, there are signs of fires along with the teeth and bones, including the charred remains of an orang-utan dated at about 35,000 years old. The implication is that human beings were eating cooked orang-utan a very long time ago.

It is known that early humans made animal traps, as well as spears and nets, with which to catch animals and to fish. What is not known is how prehistoric people hunted animals like orang-utans, which have the advantage over humans of living in trees and being able to escape through the tree canopies. Perhaps bows and arrows had been invented in those early times. Or perhaps prehistoric orang-utans travelled more often on ground, where they would have been much easier to catch. This is conceivable, because fossil teeth indicate that prehistoric orang-utans were larger than those of today, and they would have found travel easier on the ground than in the trees. Furthermore, during the Ice Ages, with their lower sea levels, mainland Asia, Sumatra and Borneo were joined, and fossil pollens suggest that there were extensive open plains at lower elevations in the region between Borneo, Java, Sumatra and Peninsular Malaysia, perhaps like the African savannah of today.

The orang-utan teeth found in Niah caves provide us with some remarkable information concerning the proportion of mothers and young. The proportion of teeth from mother orang-utans is greater than would be found in a population of wild orang-utans, while the proportion of baby teeth is much lower. From this, we can guess that prehistoric humans ate the unfortunate mothers and kept the babies as pets, perhaps allowing them out of the caves to play and grow.

The teeth of prehistoric orang-utans have also been found in caves in the Padang Highlands of West Sumatra, Indonesia. But there are curious differences from those found in Sarawak's Niah Caves. No clear evidence of human hunting activity has been found in the caves of the Padang Highlands, and most of the teeth come from adult male orang-utans. One possible explanation is that old male orang-utans went to the caves deliberately to die. We know from occasional sightings in Borneo that some mature male orang-utans travel great distances into hill ranges, away from the breeding populations. But that cannot be the sole explanation, because there are also a few baby orang-utan teeth in the Sumatran caves. An alternative interpretation is that people did indeed hunt orang-utans in Padang Highlands in prehistoric times, but that they cooked and ate them outside the caves, and kept the skulls of large orang-utans in the caves. In support of that idea we know that, until recent times, some native people of the Mentawei Islands off the coast of the West Sumatra, as well as in parts of Sarawak, hunted monkeys and apes for food, and kept the skulls hanging inside their houses. As with most forms of decoration, big ones tend to be favoured over small ones of the same form.

Orang-utans naturally attracted the attention of the early European explorers in Borneo and Sumatra. An ape referred to as Orange-autang, or Man of the Woods, is described by Nicholas Tulp in a book on medicine published in 1631. There is uncertainty as to whether this refers to a chimpanzee from Angola in Africa, or an orang-utan from Angkola in Sumatra.

Opposite (top) *The main human food of recent millennia in Borneo has been rice, grown through most parts of the island using a fallow system on non-irrigated land; forest has to be cut and burned to provide space, light and fertilizing ash before planting out the rice seeds.*

Opposite (bottom) *Cutting down forest requires metal axes, which in the past limited human spread and activities through tropical rain forests.*

tolerable good Faces (handsomer I am sure than some Hottentots that I have seen), large Teeth, no Tails nor Hair, but on those Parts where it grows on human Bodies; they are very nimble footed and mighty strong; they throw great Stones, Sticks, and Billets at those Persons who offended them. The natives do really believe that these were formally Men, but Metamorphosed into Beasts for their Blasphemy.' Captain Beeckman claimed to have purchased a young one:' I bought one out of curiosity, for six Spanish dollars; it lived with me seven months, but then died of a Flux; he was too young to show me any pranks, therefore I shall only tell you that he was a great Thief, and loved strong Liquors; for if our backs were turned, he would be at the Punch-bowl, and very often would open the Brandy Case, take out a bottle, drink plentifully, and put it very carefully into its place again. He slept lying along in a human Posture and one Hand under his Head. He could not swim, but I know not whether he might be capable of being taught.'

In the 17th and 18th centuries, most authors had never seen even a part of a non-human great ape, and there remained confusion during that period over the number, identity and origin of the living great ape species. For that reason, the Swedish naturalist Carl Linnaeus, inventor of the internationally-accepted system of scientific naming of plant and animal species, got his facts about orang-utans dreadfully wrong. In his *System of Nature* published in 1758 he wrote the following about *Homo sylvestris orang-outang*: 'It lives within the boundaries of Ethiopia, in caves of Java, Ambonia, Ternate. Body white, walks erect, less than half our size. Hair white, frizzled. Eyes orbicular; iris and pupils golden. Vision lateral, nocturnal. Life-span twenty-five years. By day hides; by night it sees, goes out, forages. Speaks in a hiss. Thinks, believes that the earth was made for it, and that sometimes it will be master again. If we may believe the travellers.'

The earliest surviving written description of what may have been an orang-utan, but may alternatively have been a hairy woman, is that of Jakob de Bondt, a Dutch doctor based in Java, who claimed to have seen a 'female satyr who hid her person with great shyness from strange men … who sighed, cried, and did a number of other actions so like the human race, that she wanted nothing of humanity but the gift of speech'. The description appears with a picture, labelled 'Ourang Outang', in a book printed posthumously in the Netherlands in 1658.

Captain Daniel Beeckman visited southern Borneo in 1712 and wrote a book about his adventures, entitled *A Voyage to and from the Island of Borneo*, in which he says: 'The monkeys, Apes, and Baboons are of many different Sorts and Shapes; but the most remarkable are those they call Oran-ootans, which in their Language signifies Men of the Woods: these grow up to be six Foot high; they walk upright, have longer Arms than Men,

Left *Captain Daniel Beeckman visited Borneo in 1712, obtaining for six Spanish dollars a young orang-utan, which died of 'flux' (dysentery) after seven months.*

Opposite (top) *Until recent decades, the difficulties associated with producing rice – availability of iron tools, adequate labour, luck with rain – greatly limited human presence in Borneo, which in turn limited human impact on wild orang-utan populations.*

Opposite (bottom) *Even before the importation of rice from mainland Asia, numans hunted orang-utans and many other wild species for food; there are signs cf 35,000-year-old cooked orang-utans at Niah caves, Sarawak.*

Linnaeus included humans in the same group as monkeys and apes, a problem for theologians of the time, and therefore for contemporary scientists as well. J.F. Blumenbach published a *Manual of Natural History* in 1779, attempting to resolve the problem by dividing primates into Quadrumana (four-handed apes and monkeys) and Bimana (two handed-humans). In the same year Peter Camper, a Dutch anatomist, published the first accurate description of the orang-utan, based on observations and dissections of several young specimens. In 1778, Arnaut Vosmaer had published pictures and a brief description of a live orang-utan in the menagerie of King William V of Orange.

As skeletons, skulls and skins became available in Europe, mainly via the Dutch presence in Borneo and Sumatra, various names were given to the orang-utan by European museum zoologists through the 19th century, including *Pongo wurmbii*, *Simia morio* and *Pithecus satyrus*.

S. Müller wrote in about 1840 that the orang-utan 'appears to be distributed over the whole of this enormous island (Borneo) with the exception of the mountains and the more populous lowlands… The broad stretches of low and level

alluvial terrain along these two coastal areas (Kahayan River and Sampit, in Central Kalimantan) … provide him with a spacious and secure abode'. In 1841, the British 'Rajah' James Brooke of Sarawak wrote 'a letter relating to the orang-utan of Borneo' published in the Proceedings of the Zoological Society of London, suggesting the existence of three species which, following local Iban terminology, he called maias papan, maias kasar and maias rambi. Subsequent writings by Alfred Russel Wallace, co-author of the original theory of evolution by natural selection, indicate that these names most probably referred respectively to adult males with large cheek flanges, adult males without cheek flanges, and adult females. In the late 19th century, however, many museum zoologists preferred to create new names for any specimens that differed slightly from previously known collections. E. Selenka in 1896 proposed several different sub-species of orang-utan, despite the fact that he had as evidence only a few specimens collected from just one part of West Kalimantan.

During the 19th century and the early years of the 20th century, several Europeans spent extended periods in parts of

Borneo and Sumatra that had previously remained unknown elsewhere, and left written records of their travels. Whether they were explorers, scientists, colonial administrators or simple adventurers, always high on the list of interesting features of their expedition was the orang-utan. In general, the 19th-century Europeans in Borneo regarded orang-utans as no different from other mammals, as they happily shot large numbers. Even Alfred Russel Wallace, a man in many ways more sensitive than his contemporaries, did not flinch from personally shooting dead sixteen orang-utans during his time in Sarawak in 1855. Thirty years later, an American collector of zoological specimens, William Hornaday, shot over forty orang-utans in the same region of swamp forest not far from the State capital of Kuching.

Of interest to the modern scientist and conservationist are indications that the natural distribution of orang-utans in Borneo and Sumatra a century and a half ago was broadly similar to what has been confirmed in recent decades. Of course, orang-utan numbers and distribution have declined, but it is clear that those early writers actually saw wild orang-utans in the general areas where the species still occurs, while those regions where

they found no trace of orang-utans during the 19th century have subsequently been proved to lack the species. For example, S. Müller reported in the 1830s that orang-utans were virtually absent from what is now South Kalimantan, whereas the species was 'less elusive' a few days' journey further westward. Hugh Low reported in 1848 that the orang-utan was not found 'in the Sarawak territory, nor in that of Sambas', but that they did occur between the Sadong and Batang Lupar Rivers. E. Selenka sought and found them in 1896 in the upper Kapuas River of West Kalimantan, while W.L. Abbott found none south of the

Opposite *In the 18th century, J. F. Blumenbach attempted to satisfy the contemporary view that humans are set uniquely apart from all other mammals by separating the primates into Quadrumana (four-handed monkeys and apes) and Bimana (two-handed humans).*

Above *S. Müller found in the 1830s that orang-utans were common in the extensive coastal swamps of Central Kalimantan. From the 1970s, numerous local and foreign visitors have come to see wild orang-utans in this same region, at Tanjung Puting National Park, entering via the small town of Kumai.*

Mahakam River during 1908–09. Based on his extensive travels through southern, central and eastern Kalimantan from 1914 to 1916, Carl Lumholtz stated that 'This intelligent, man-like ape is probably not so common in Dutch Borneo (Kalimantan) as he is supposed to be. Mr Harry C. Raven, who collected animals in the north-eastern part, told me that in a year he had shot only one. The orang-utans are generally found in Southern Borneo and do not go very far inland; in Central Borneo they are extremely rare, almost unknown.'

Seemingly, confusion sometimes arose among local inhabitants and foreigners alike as to whether they were looking for apes, remote tribes or forest spirits. When the English zoological collector John Whitehead was on Mount Kinabalu in 1888, one of his local assistants returned to the camp in a fright, having seen a hantu (ghost) peering down at him from a tree. Whitehead, who did not witness this apparition, wrote in his journal that this must have been an orang-utan.

While some of the native people whom these European interlopers met in Borneo and Sumatra shared a common relish for slaughtering orang-utans (for food or decoration, rather than for science or fun), some communities, or some individuals within a community, had and still have taboos against harming these apes. For example, the survival until modern times of orang-utans in Lanjak-Entimau Wildlife Sanctuary is attributed in large part to the fact that some local Iban communities have a long-standing taboo against hunting and eating this species.

While no single orang-utan was ever able to get his own back on his human tormentors, who often excused their acts as science, the species as a whole managed to do so in a very strange way. This happened because an orang-utan featured in

one of the greatest scientific hoaxes of the early 20th century. In 1912, an amateur archaeologist named Charles Dawson reported finding the remains of prehistoric man in a gravel excavation pit in southern England. Dawson produced parts of a very old skull and jawbone of something that appeared to be intermediate between a human and an ape. The find caused a sensation. Here at last seemed to be proof of 'the missing link' between apes and humans, sought for decades by those, scientist and amateur alike, who were fascinated by the Theory of Evolution by Natural Selection, first proposed more than half a century earlier by Charles Darwin and Alfred Wallace. There were a number of sceptics who found it difficult to believe that the remains came from a place where no similar fossils had been found, of all places right in the heart of the British Empire. But it was not until 1955 that careful examination showed that the skull was that of a modern human, albeit probably hundreds of years old, while the jawbone was that of an orang-utan. It is believed that the orang-utan bone was also hundreds of years old, and that it came from gold washings of early Chinese settlers in Sarawak.

Above (left) *European explorers of the colonial era noted that orang-utans were most common in southern Borneo; this old male is in Tanjung Puting National Park.*

Above (right) *Making zoological collections on Mount Kinabalu in 1888, John Whitehead's assistant reported having seen a ghost peering at him from a tree, which Whitehead assumed to be an orang-utan.*

Opposite *A picture from the first edition of* The Malay Archipelago, the land of the orang-utan, and the bird of paradise, *by Alfred Russel Wallace, published in 1869, which bears the caption, 'Orang-utan attacked by Dayaks'.*

Myth and folk tales

All the large apes have long held a special fascination for humans. Their combination of wildness, intelligence and great physical strength induces a peculiar hold on the imagination. Lurking at the back of our minds also is the notion that wild male apes may have a desire for beautiful human females. These notions were exaggerated into the 1933 film *King Kong*. In appearance, King Kong resembles an African gorilla, but his home was supposedly an island off Java, in Indonesia, implying a relationship with the orang-utan. This piece of artistic licence helped to confuse millions and also perpetuated vague ideas in many minds of the existence of gigantic wild apes somewhere in the general vicinity of Indonesia. Occasional tales surface of wild orang-utans attacking or molesting humans, but none has ever been proved to be true. Not surprisingly, however, if attacked by humans, as has happened in some plantations in recent decades, orang-utans may retaliate.

There are various traditional stories and beliefs involving orang-utans in some of the regions of their origin. Some beliefs are widespread in Borneo and may date from very early times. For some people in central Borneo, it is considered that to look into the face of an orang-utan is to invite bad luck. Laughing at

its face is even worse. In fact, laughing at or ridiculing animals is traditionally regarded by some South-east Asian rural societies as shocking behaviour, likely to endanger the safety of the whole community or to cause the person involved to turn to stone. And some communities believe that a pregnant woman should never look at an orang-utan. There are also stories that tell how, in days gone by, orang-utans could talk, and that they stopped talking for fear of being enslaved by human beings. There are several variations on a story involving a mixed marriage between a human being and an orang-utan, resulting in the birth of a child. Some stories involve a girl abducted by a male orang-utan, while in others a man hunting in the forest is abducted by a female orang-utan.

Left *Several myths are associated with orang-utans, including one that the species stopped talking many generations ago for fear of being enslaved by humans…*

Above *…and that laughing at their face invites bad luck.*

Opposite (top) *Large apes have long held a special fascination for humans – their combination of wildness, intelligence and great physical strength induces a peculiar hold on the imagination.*

Opposite (bottom) *Occasional tales surface of wild orang-utans attacking humans, but – with the exception of orang-utans cornered and provoked in plantations – not one is true.*

The following story, recorded by Kinabalu Park Naturalist Ansow Gunsalam and related it detail in the 1983 issue of the Sabah Society Journal, is one of the most detailed versions. In this story, a hunter from the Pinosuk area of the slopes of Mount Kinabalu was out in the forest when he came upon a female orang-utan. In those days, orang-utans were thought not only to be able to talk, but also to recite magic words which would protect them from hunters. When such orang-utans looked directly at the hunter, no object could pierce their skin. This female orang-utan came to the hunter and signalled to him to ride on her back to her nest in a tall tree near the Liwagu River. He found the nest well stocked with fruits, with which the orang-utan would feed him when he was hungry. After two years of this life, the orang-utan became pregnant, although the hunter was still trying to think of ways to escape. One idea that he had was to suggest that the orang-utan obtain a long rattan cane, on the pretext of lining the nest for the coming baby, but actually to enable him to climb down to the ground. After the rattan was in place, he suggested that the orang-utan go out to stock up with food for the new baby, thus giving him enough time to escape. But she was never away for a long enough time. Then the hunter insisted that she must go to a village to obtain rice, which the baby would need. The hunter started to escape, but just as he reached the ground, the orang-utan returned to the nest. Without being seen, he rushed and dived into the Liwagu River. The river was flooding, and he was swept downstream. He managed to grasp on to a rock and, leaving his belongings, including a quiver of poisoned darts, on a sandbank, hid in a nearby rock hole. The orang-utan found the belongings and reached into the rock hole but, being too large from pregnancy, she could neither feel nor see him inside. In despair at losing her husband, she stabbed herself with the poisoned darts, crying and shrieking. When the noises stopped, the hunter emerged and found the orang-utan dead. He cut open her belly, and removed a live baby boy, which looked like a human being but was covered with hair. The hunter buried the dead orang-utan and returned to his village with the newborn baby. There, he found only one old man remaining, along with old and young women, including his true wife. Having related his story, the

Left *Legends in Sabah and elsewhere in Borneo tell of native hunters who were seduced and kept by a wild female orang-utan.*

hunter was told that a head-hunting tribe had raided the village and killed or taken all males except the old man. The head-hunters threatened to return in ten years' time. The hunter raised his hairy child as a warrior to destroy these head-hunting enemies. Within ten years, the head-hunters returned, and in front of their womenfolk performed a dance on the bank of Liwagu River, the bachelors promising heads as a sign of bravery. The hairy youth became infuriated and, despite pleas from his father to arm himself, strode into the enemy group and in a fury tore every one of them apart with his bare hands. His proud father instructed him to select any girl as a wife, but instead, in the following weeks, the son took to spending longer and longer periods away from the village. Finally, he disappeared and was never seen or heard of again.

In modern times, a number of human thinkers do not take too seriously the DNA story that chimpanzees are closer to humans than orang-utans, while many are not convinced that the gap between humans and other great apes is as wide as most of us assume. For example, J.R. Grehan of Buffalo Museum, New York, compares great ape traits other than DNA and hard anatomy (such as concealed ovulation, male beard, prolonged mating, 'house' construction, mechanical 'genius' and artistic expression) and suggests that orang-utans and humans have more in common than do the other living apes.

Many researchers in the social and anthropological fields have observed orang-utans through qualitative and quantitative studies, and concluded that orang-utans have a variety of traits and characteristics that indicate that humans are not totally distinct from the animal kingdom. The ability to use tools, to learn and understand the meaning of spoken human words, and to paint pictures are all interesting abilities, but do not show that orang-utans are extra special, as some birds, dolphins and elephants possess similar abilities. But orang-utans also have some ability to use syntax, and to show awareness of 'self', while populations of orang-utans develop elements of culture. Based on these and other considerations, one group, Great Ape Project, has published a charter declaring that all great apes should enjoy the right to freedom from being killed or imprisoned.

COMMUNICATION AND CULTURE

The orang-utan is one of the most intelligent living species – arguably the most intelligent after human beings. Researchers of wild orang-utans, as well as observers of zoo orang-utans, have always been especially interested in how these creatures communicate with each other, and how they might be able to communicate with humans.

In their pioneering study of wild orang-utans, Herman Rijksen and John MacKinnon recorded nearly fifty distinct types of behaviour involving one orang-utan interacting in some way with another orang-utan so as to convey information. The list ranged from aggressive behaviours such as hitting, to clinging, nodding, watching, hiding, temper tantrums, mothers using their body as a bridge for infants, to behaviours associated with mating. These researchers also distinguished more than twelve kinds of vocalization (noises made through the mouth). An example is the 'kiss-squeak', in which air is sucked in between pursed lips when an orang-utan is slightly irritated by something, such as noticing a human observer on the ground. Long calls made by flanged males are not only a distinctive feature of the species, but for humans one of the most thrilling and memorable sounds of the Bornean and Sumatran rainforests. These calls – a series of organized groans and roars, lasting between about one to three minutes – start with a soft, bubbly introduction, followed by deep, loud roars and ending with sighs.

Researchers are also interested to know whether wild orang-utans have distinctive behaviours over and above those required for basic survival needs, and which differ between different areas. Such behaviours would indicate that the species has developed a culture.

When researchers who had observed wild orang-utans at all the different major sites in Borneo and Sumatra compared notes, it was found that 17 behaviour variants were found to be common in at least one site, but absent at another site, despite there being no obvious possible ecological reasons for such differences. Some of these cultural differences were limited to either Borneo or Sumatra. For example, 'roof nests' (a nest built above the sleeping nest, to shelter the occupant from rain) and 'kiss-squeak with leaves' (giving a typical kiss-squeak while holding leaves in front of the mouth) were seen only in Borneo, while 'tool use for tree holes' (using a twig to obtain insects or honey from tree holes) and 'branch scoop' (using a small leafy branch to scoop drinking water from tree holes) were seen only in

Opposite *A common vocalization among all wild orang-utans is the 'kiss-squeak', made when the animal is irritated by something – such as seeing a human observer.*

Sumatra. But there are also some behaviours which occur in some orang-utan populations on both Borneo and Sumatra, but are also absent from other populations on both islands. Examples of such behaviours are 'nest raspberries' (making a 'raspberry' noise through pursed lips as nest building is completed) and 'sun covers' (piling leaves or branches over the body or the nest on sunny days).

Overall, wild Sumatran orang-utans have a wider repertoire of cultural behaviours. To date, wild orang-utans in Sumatra have been seen to use tools, such as twigs to obtain insects from holes and leaves to hold spiny fruits, while wild Bornean orang-utans have not been seen to use such tools. For example, Sumatran orang-utans throughout the Suaq Balimbing swamp forest use short sticks on *Neesia* fruits, which contain masses of tiny irritant hairs that protect the seeds, in order to dislodge and eat the seeds. But in zoos, Bornean orang-utans adopt a variety of tool-using behaviour, which may then be copied by others. In fact, all captive orang-utans can quickly learn tool use from humans and from other orang-utans, surpassing chimpanzees not only in learning, but in creating their own inventions. At Camp Leakey, Tanjung Puting National Park in Central Kalimantan, orang-utans under rehabilitation from captivity have been seen to imitate such actions as 'washing' clothes, 'brushing' teeth, 'sawing' logs, 'hammering' nails and voluntarily riding in boats.

By comparing all the studies of wild orang-utans, it is clear that, overall, Sumatran orang-utans are more sociable than their Bornean relatives. This conclusion can be drawn from general observation, and is confirmed by quantitative measures, such as how much time two wild orang-utans spend 50, 10 and 2 metres from each other. From these results, researchers believe that the use of tools is linked to social behaviour. The less sociable Bornean orang-utans learn the majority of their behaviour just from their mother, while Sumatran orang-utans have more opportunity to copy from a wider array of other orang-utans.

In captivity, orang-utans can be given the opportunity to do things that would be virtually impossible in the wild. One is to make paintings with a variety of colours. Another is to learn a language, based on the human concepts of assigning specific words to specific objects or situations, as well as describing particular feelings. Although orang-utans cannot speak in any way that resembles human speech, they can learn sign language. Psychological research has shown that orang-utans, like other great apes, have a consciousness of 'self'.

Above *Orang-utans are intelligent and inquisitive, learning new behaviours from other orang-utans and from humans.*

Left *This female orang-utan at Bohorok, Sumatra, solicits regular boat rides, apparently for enjoyment.*

Opposite *Some orang-utans in some wild populations in both Borneo and Sumatra make leaf umbrellas as protection from rain or sun.*

HEALTH AND DISEASE

Orang-utans suffer from many of the diseases and health problems that are experienced by humans, particularly those affecting the respiratory and digestive tracts. Symptoms resembling colds and flu are quite often noticed in orang-utans in the wild, in rehabilitation centres and in zoos, especially during wet weather. Any kind of stress or poor conditions may increase susceptibility to illness, and this may occur with orang-utans in natural forests, rehabilitation centres and zoos. The existence of the orang-utan rehabilitation centre at Sepilok, Sabah, and subsequently similar sites in Indonesia, where recently captured wild orang-utans have been available for examination since 1965, has had the unintended benefit of yielding much information on orang-utan health and disease.

There are species of disease-causing organism which have bad effects on humans but not on orang-utans, and vice versa. But some of the most debilitating diseases shared by orang-utans and humans are caused by exactly the same organisms. Bacterial diseases which may affect orang-utans in contact with humans include tuberculosis (caused by *Mycobacterium tuberculosis*), dysentery or diarrhoea (caused by *Shigella*, *Salmonella* and *Campylobacter* bacteria species, which may occur in food or water), and meliodosis (caused by *Burkholderia pseudomallei*, which lives in soil and stagnant water – infections may result in fevers, abscesses and septic shock) are all recognized as threats to orang-utans at rehabilitation centres and occasionally in zoos. Orang-utans and humans also suffer respiratory problems from the bacteria *Klebsiella pneumoniae* and *Enterobacter* species.

To some extent, zoo orang-utans are shielded from risk, because they are cleaned and well cared-for by their keepers, have a good diet and access to medical care. Also, since all zoo orang-utans are either captive bred or have been in the zoo for decades, they have had the opportunity to develop some degree of resistance to human pathogens. Orang-utans in rehabilitation centres in Malaysia and Indonesia, on the other hand, may be at greater risk because, even though they may come into contact with fewer people and have medical care, they tend to be young and to suffer from a background of poor diet and stress. They often travel and play on the ground, where it is impossible to maintain the degree of cleanliness that exists either in a truly wild forest environment or on scrubbed hard floors. Wild orang-utans eat a very wide variety of plant species and plant parts, sometimes feeding on small amounts of a plant which they normally ignore. Some primatologists have speculated that such plants may have medicinal properties, thereby suggesting another possible reason why wild orang-utans tend to enjoy better health than their rehabilitant counterparts.

Evidence of a wide range of viral infections has been found in wild orang-utans, as well as in orang-utans in rehabilitation centres in Borneo, and in zoo orang-utans outside their natural range areas. William Karesh, N.D. Wolfe, Annalisa Kilbourn and their colleagues of Wildlife Conservation Society, Johns Hopkins School of Hygiene and Public Health and Sabah Wildlife Department conducted a study of 84 wild orang-utans (captured

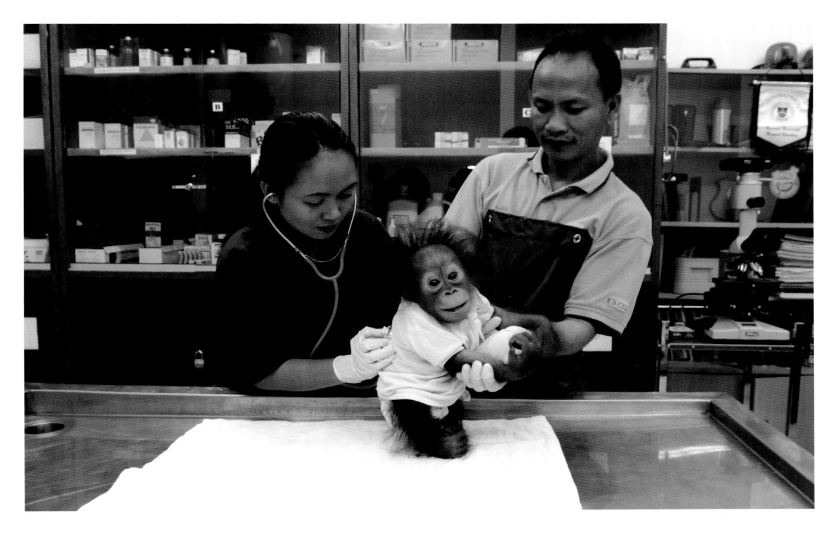

from areas being converted to plantations, for relocation to protected areas) and 60 orang-utans at Sepilok rehabilitation centre (see Further Reading on page 154). They found evidence of 47 types of virus. Common viruses in the wild orang-utans included Japanese encephalitis and foamy virus (a benign retrovirus), while parainfluenza (causing respiratory tract problems) and langat viruses (transmitted by ticks) were found only in those at Sepilok. Dengue and mumps viruses were found in a few orang-utans in both wild and rehabilitant groups. In zoo orang-utans with sickness caused by viruses, the infections are most likely to have come from humans or from other primate species in the zoo. There are recorded cases of zoo orang-utans where viral infection by coxsackie virus, *Herpes simplex* or Hepatitis B virus was the main or sole cause of death. The various *Herpes* viruses tend to be harmless in their natural host animal species but fatal in other animal species. For example, *Herpes simiae* can be fatal to humans, while *Herpes simplex* can be fatal to orang-utans.

Among protozoan parasites, the ciliate protozoon *Balantidium coli*, which is ingested and causes dysentery, has been found in wild orang-utans and rehabilitant orang-utans in Sabah. Another protozoan, the encephalitis-causing *Balamuthia mandrillaris*, first identified in zoo mandrills, is of uncertain origin in the wild but has led to the death of zoo orang-utans.

Pioneering studies at Sepilok in the early 1970s showed that many of the orang-utans there have two species of malaria parasite (*Plasmodium* species) which do not infect humans, while the orang-utans are free of the three malaria species that cause illness and sometimes death in humans. Orang-utans with malaria suffer occasional fevers and weakness, but a natural partial immunity seems to prevent more drastic effects. Studies have found that only about 12 per cent of wild orang-utans have been infected with malaria while over 90 per cent of rehabilitant individuals suffer from the disease. It is believed that the constant proximity of many orang-utans in one place accounts for this difference. The finding is interesting in showing how the wild orang-utans' solitary nature may play a role in keeping them free of this and other diseases. Another intriguing recent finding is that orang-utans may have natural genetic resistance to the bad effects of malaria, but that the gene that confers this resistance is also linked to thalassaemia.

Opposite (top) *The existence of the rehabilitation centre at Sepilok since the 1960s has contributed to our information on orang-utan health and disease, including the fact that some diseases are shared by humans and orang-utans.*

Opposite (bottom) *Wild orang-utans feed on a wide variety of plants, sometimes eating small amounts of a species normally ignored, leading to speculation that the latter may have medicinal properties.*

Above *Some orang-utans cared for at Sepilok contract diseases, especially when young and before they have developed resistance; the clinic provides diagnosis, care and treatment.*

Above *Orang-utans that frequently walk on the ground are at greater risk of contracting parasites than those that are almost entirely tree-dwelling.*

Tiny parasitic worms are often detected in wild, rehabilitant and zoo orang-utans. The most common infective nematode found in primates is *Strongyloides*, of which three species causing serious illness are known. The nematode larvae, which occur free-living or in faeces, penetrate skin and migrate via the blood to the respiratory organs. They may cause pneumonia and, if coughed and swallowed, enteritis. Liver flukes (*Platynosomum fastosum*) have been found after death in formerly-captive orang-utans brought to a rehabilitation centre in East Kalimantan. It is believed that the source of this parasite was domestic cats.

A combination at one time of more than one of the above diseases in one orang-utan may be very debilitating. We know that influenza may be transmitted over long distances by wild birds, which carry the virus without ill effects. It is possible that an epidemic of influenza, acting on wild orang-utans with malaria or other parasites, might wipe out whole populations over large areas. This is a possible contributory reason for the absence of orang-utans from some regions.

Non-infectious health problems probably differ between wild and zoo orang-utans. Researchers occasionally witness or infer falls of orang-utans from trees. Herman Rijksen recorded seeing at least twelve falls within three years, one from a height of 30 metres, with one death resulting from the fall, and another inferred. 'Old-age' diseases are not recorded from wild orang-utans, but cardiovascular disease, diabetes, malignant tumours and inflammatory arthritis have all been recorded in some older zoo orang-utans.

NAMES FOR ORANG-UTANS

The name orang-utan is clearly derived from the Malay word *orang hutan*, which used to be translated as 'man of the forest', although equally correct is 'forest people'. It is very unlikely, however, that this name was originally used by native people in Malaysia or Indonesia to refer to the species which nowadays is known everywhere as orang-utan. Quite possibly the term was an invention coined for convenience between local residents and early European explorers. The earliest written records date from the first half of the 17th century, when variants on the spelling *orang-outang* were used. This spelling is still used in some zoos.

In Sabah, native Kadazan and Dusun peoples call orang-utans *kogiu* (pronounced 'cog you'), while Murut people in the Sabah-Kalimantan border area know them as *kahui,* and Orang Sungai in the eastern parts of Sabah use either *kogiu* or *kisau*. Throughout most of Kalimantan, orang-utans have various but similar-sounding local names, including *kehiau, kahiu, keau, kagiu, kayies* and *kayas*. The Lawangan Dayaks, original inhabitants of the border area between South and Central Kalimantan, use the name *juwud*. In Sarawak, orang-utans are called *maias*, which is remarkably similar to *mawas,* the name used in northern Sumatra.

Some rural people in Borneo say that there are three kinds of orang-utan. In Sarawak, they may be known as: *maias kesa,* which corresponds to smaller individuals; *maias rambai,* which may represent both mature females and males which are not fully mature; and *maias timbau,* or *maias papan,* which is the fully mature male. This kind of naming is not unusual, as Borneo natives also give special names to wild deer and pigs of different size and sex.

Scientists now recognize only one or two species of orang-utan, with the commonly accepted genus name of *Pongo*. Unfortunately, this name seems to have been based on an uncertainty among early zoologists, who were not sure of the origins or relationships of either African apes or orang-utans, and who borrowed *mpongo*, a Congolese name for the gorilla. A few zoologists prefer, therefore, the use of an old alternative scientific name for the orang-utan, *Simia satyrus*, but this is not generally recognized.

Opposite *The local name for the orang-utan in Sumatra is* mawas, *while a wide range of indigenous names are used in Borneo; the term 'orang-utan' (meaning 'forest people') was probably coined for convenience in discussion between local residents and early foreign explorers.*

CHAPTER FOUR

WILD ORANG-UTANS

ORANG-UTANS ARE A CURIOUS MIX. ON THE ONE HAND, THEY ARE FRUIT-EATING SPECIALISTS THAT REPRODUCE VERY SLOWLY IN BIODIVERSE FORESTS. ON THE OTHER, INDIVIDUAL ORANG-UTANS SEEM ABLE TO TOLERATE EXTREME PRIVATION, LIVING FOR YEARS ON A FAMINE DIET IN DEGRADED OR INFERTILE HABITS. THESE CHARACTERISTICS SUGGEST THAT WILD POPULATIONS COULD REACH A TIPPING POINT, WHEREBY CHRONIC POOR NUTRITION, STRESS AND A SUBTLE TREND OF INCREASING MORTALITY MAY COMBINE TO CAUSE RAPID LOCAL EXTINCTION.

DISTRIBUTION, RANGE SIZE AND POPULATION DENSITY

The species distribution, individual home range and population density of wild orang-utans are three factors that have an important bearing on our understanding of their biology and ecology, and how we can try to address their conservation needs.

Distribution

Distribution means the overall range of the species. The map on page 17 shows the approximate current distribution of the Sumatran orang-utan and of the three sub-species of the Bornean orang-utan. There are rarely sharp cut-off lines between areas where orang-utans occur and where they do not. Generally, though, we can recognize clear areas where there are constant breeding populations, more fuzzy-edged zones where some orang-utans spend some of their time, and vast regions where there are normally no orang-utans at all. In these latter areas, mature solitary male orang-utans roam and are occasionally seen. It is believed that these are individuals that have moved far from their birth place and out of the zones occupied by breeding populations. In 1960, for

example, Tom Harrisson — curator of the Sarawak Museum and husband of Barbara — published in the Sarawak Museum Journal a paper entitled 'A remarkably remote orang-utan', in which he described the killing of an unusually large male orang-utan in the upper Baram River, well away from the known main orang-utan populations of Borneo. Harrisson described the arm bones as being up to one inch longer and thicker than those of other male orang-utan arm bones known to him.

For people concerned about the future survival of the orang-utan in the wild, it is important that most attention is given to those areas where there are quite large breeding populations. Clear proof of the locations of areas where orang-utans can survive and breed is a more basic issue than concern over the difficulties involved in securing permanent protection for the land and forest in those areas, however challenging that might be. The same cannot be said for forests where there are either no, or very few, or only occasional wandering individuals. Perhaps those forests are in some way unsuitable for orang-utans. But that does raise the question why there were huge forest areas lacking any orang-utans in recent historical times, even before widespread forest loss.

The most favoured explanation is that human hunters have gradually wiped out the species over thousands of years, right up to the present. There is no doubt that this has happened in many places. The evidence of human hunting of orang-utans over many thousands of years at Niah in Sarawak, and the species' current absence there, is as close as one can get to proof that sustained hunting can cause the extinction of orang-utans, however good the habitat. And the absence of orang-utans throughout most of Sumatra, with its fruit-rich forests, even before the arrival of the Dutch, can hardly be put down to unsuitability of the habitat. A study of over one hundred sites in East Kalimantan showed that the presence of orang-utan nests was not related to any aspect of forest ecology or to human presence, but was statistically correlated to the distance from the nearest village where residents were known to hunt orang-utans.

But hunting cannot be the sole explanation for the limited and patchy distribution of orang-utans. Consider the fact that the highest population densities are in swamps and fertile valley and lowland soils, and that there are vast areas of inland hill forests with no orang-utans and no evidence of human hunting, past or recent. Generally, orang-utans are absent from the least fertile soils, such as extensive hill and mountain ranges, and from heath forests. This makes good ecological sense. Over long periods, there will be the smallest quantities of fruits on the least fertile soils. We have seen already that, for hundreds of years, occasional long droughts have occurred in Borneo and Sumatra. Any one of those might have led to large numbers of orang-utans starving to death, especially in forests that tend to dry out the most during those periods. Disease, or combinations of diseases arising from weather or food stress and pure luck, may also have caused stress on affected orang-utan populations, sufficient to wipe them out in the worst cases.

Previous pages *Orang-utans are fruit eating specialists with a very low reproductive rate.*

Opposite *Mature male orang-utans may range into forested hills and mountains, far outside the limits of breeding populations.*

Above *There are vast areas of inland hill forests in Borneo and Sumatra with no orang-utans, and no evidence of human hunting past or recent.*

Range size

Range size (individual home range, to give the correct term) refers to the total area used by an individual orang-utan during its lifetime following departure from its mother. Naturally, like most other animals as well as humans, some individuals have much bigger ranges than others, while all spend more time in certain places while very rarely visiting other places. And of course, the longer one observes a particular orang-utan, the bigger its range size appears to become. There is no special reason why orang-utans might cover their entire range within just one year. But there is no consensus among researchers or the usual range size for orang-utans.

Researchers in Sabah, Kalimantan and Sumatra from the 1960s to the 1990s generally suggested that each orang-utan used an area of between about 50 to several hundreds of hectares. Some were not sure whether ranges might be more extensive, because their overall study areas were generally no more than a few hundred hectares. Most felt that the ranges of males were larger than those of females. All agreed that individual orang-utan home ranges overlapped with the ranges of many other orang-utans. Generally, ranges of Bornean orang-utans seemed to be rather smaller than those of Sumatran orang-utans. Some researchers suggested that there were some individual orang-utans with ranges of tens to hundreds of hectares, but that some other individuals were nomadic over large areas.

To try to account for the different interpretations of different researchers, and the fact that all researchers agreed that some orang-utans were seen once or a few times only, and then never again, Herman Rijksen and Erik Meijaard in 1999 suggested that orang-utans might be divided into three classes: residents, which live in relatively small ranges of up to a few hundred hectares; 'commuters', which stay in one area for a while, disappear, and are then seen again from time to time in following years; and 'wanderers', which are seen once only, or once every few years, and may range over huge areas.

Following research in the coastal swamp forests at Suaq Balimbing, Sumatra, between 1996 and 1998, Ian Singleton concluded that most orang-utans there – and by implication in at least some other areas – have much larger individual ranges. He suggested that female home ranges averaged between 900 and 1,500 hectares, with a core use area of about 500 hectares, and that adult male ranges were 3,000 to 10,000 hectares.

If we combine the results and thinking of the different researchers, the most likely explanation seems to be that different individual orang-utans end up occupying vastly different sizes of home range. Some spend their life in small ranges and can be found quite easily by human researchers, while others use very much larger areas, concentrating on parts of their range for a while and then moving elsewhere for extended periods. The orang-utans' great intelligence in terms of remembering past food sources, and being able to guess when they might be available again, allow them to choose small or large home ranges. The strategy of sticking with a small range restricts the individual to lots of small food sources, many of them not particularly favoured, such as leaves and fruits of small trees and lianas. The strategy of a large range means depending on variations in fruiting peaks among different tree species in different localities, but the reward of large amounts of preferred fruits at most times. The strategy of very large ranges, adopted only by some males, confers the opportunity to mate with many females, the risk of fathering few or no children, and the need to simply feed wherever food sources are found. Increasingly, as protected areas containing orang-utans become isolated by changes in surrounding land use,

the surviving orang-utans will have little choice as to their individual home range size. Sepilok Forest Reserve in Sabah is only 4,300 hectares in extent. The orang-utans in the lower Kinabatangan are in an even more precarious position, with many separate forest blocks, and some individuals now sharing just a few hundred up to a couple of thousand hectares with a few tens or a couple of hundred other orang-utans. To ensure the survival of this population, there will need to be long-term efforts to re-join the separate forest blocks and to occasionally move selected orang-utans from one block to another, so as to minimize the risks of inbreeding. The intensive work of the Kinabatangan Orang-utan Conservation Project with Michael Bruford and Benoît Goossens of Cardiff University, identifying individual orang-utans and their relationships through DNA analysis, has provided specific suggestions on the best approaches towards addressing the long-term inbreeding problem. There is another concern, however, that the separate habitat blocks may be too small to provide sufficient food for all the orang-utans during certain periods.

Opposite *The extent of the total home range of individual orang-utans varies enormously between individuals, habitats and geographical locations, from a few tens to thousands of hectares (note the orang-utan moving though the tree canopy on the right side of this picture).*

Above *As habitats containing orang-utans become isolated by changes in surrounding land use, resident orang-utans will face habitat limitations not experienced by their ancestors in the days before the expansion of human activities. Sepilok Forest Reserve (4,300 hectares, shown here) was established as a protected area in 1957.*

Population density

Population density is simply a measure of how dense the orang-utan population is in any given area, expressed as the number of orang-utans per square kilometre. The value of this measure is to gauge the suitability of the forest area in terms of supporting orang-utan populations. If a study shows that there is a high average population density of orange-utans in a particular area (say, more than two orang-utans per square kilometre), and that the overall forest area is large, we know that the area is of great significance for conserving the species, and that efforts must be made to conserve that forest. If only a small area has a high population density of orang-utans, the reason might be that there is not much forest left, and that orang-utans have been squeezed in by conversion to plantations. Consideration may have to be given to a properly planned rescue operation to transfer these individuals to a large protected area. But if there is a good chance to preserve the forest, there is a case to formulate a plan for a new protected area. In the longer term, there will need to be options to restore forest or enlarge the extent of habitat, or link the area to another forest by replanting along a river valley, or occasionally transferring selected orang-utans between forest blocks so as to retain high genetic diversity and prevent inbreeding.

In areas where there are few orang-utans, the population density concept becomes rather abstract. For example, we may do a survey and obtain a result of 0.2 orang-utans per square kilometre, which can also be expressed as one orang-utan per five square kilometres. If there has been no ongoing or historically recent hunting of orang-utans in such areas, we need to consider the possibility that the forest might not sustain breeding populations of orang-utans in the absence of adjacent lowland forest. In turn, that indicates doubt over the long-term suitability of the area for orang-utan conservation, especially if the nearby lowlands are to be converted to plantations.

Right (top) *Average population densities of orang-utans vary between forest types – in primary Borneo dipterocarp far from the coast (as shown here), there are usually few or no orang-utans.*

Right (bottom) *Dipterocarp forests in Sumatra may support several orang-utans per square kilometre.*

Opposite *A rare picture of orang-utans mating, in this case an unflanged male forcing himself on a biologically receptive but unwilling adult female.*

Left   *Orang-utans feed on both ripe and unripe fruits (including durians, as shown here).*

Below   *Fruits of trees of the Meliaceae (mahogany) family represent a common food of Bornean orang-utans throughout the species' range.*

Opposite   *Fruits of the riverside tangkol fig tree (Ficus racemosa) are another staple food of the Kinabatangan orang-utans.*

THE ORANG-UTAN'S DIET

Looked at through the scientific eyes of a zoologist, the essence of an animal is what it eats. Put simply, orang-utans eat fruit. All researchers who have studied wild orang-utans agree that seeking food and eating represent the activities that use up most of the ape's time and effort. And that the kind of food that orang-utans prefer is fruit, of almost any kind, but particularly those fruits that are rich in carbohydrates and protein.

Fruits in tropical rainforests are not like fruits in temperate forests or in plantations, in the sense of large amounts of just a few kinds being available seasonally. But nor do fruits in tropical rainforests come in a constant abundant supply of many juicy varieties. Tropical rainforests have strong seasonality in flowering and fruiting, based largely on the occurrence of very dry and very wet periods, which vary from year to year in spacing and intensity. Exact patterns and amounts of rainfall vary to some extent from valley to valley, and hill range to hill range. Coupled with differences in soils and other factors, that means that peaks and troughs in fruiting are different between localities over wide areas, while within a particular locality there may be periods with very little fruit interspersed with unpredictable peaks.

The form, nutritional value, palatability and crop size of fruits vary enormously between different plant species. Some strangler figs may produce millions of palatable fruits while there are no other figs fruiting nearby. Some tree species which happen to be locally common throughout an entire valley may produce lots of fruits simultaneously every two or three years. Other species may produce repeated smaller crops throughout the year, every year. Lianas may produce small, scattered crops of fruit. The majority of tropical fruit species are rather small, tough, fibrous, bitter things, even when ripe.

Fruit-eating bats (such as flying foxes) and birds (such as hornbills and pigeons) have evolved to make use of their wings, to fly over large distances seeking their preferred ripe fruits. Orang-utans, relying on slowly climbing through the forest canopy, are faced with a much greater challenge. They have to simultaneously use their intelligence and experience to track down today's and next week's meals, and to be content with almost any type of fruit, ripe or unripe, bitter or sweet.

There are some zoologists who believe that the orang-utan's diet, coupled with the challenges of finding enough fruits year round, are the key to understanding almost all other aspects of the species' biology. The argument goes that high intelligence and reasoning skills are necessary attributes in coping with the limited travel options available in rainforest to a large, non-flying animal, coupled with constant uncertainty over fruit supplies. Orang-utans need to be able to remember all the subtle differences in the fruiting patterns of different tree species, as well as where individual trees are and were fruiting a few years ago. They also need the ability to decide whether it makes better sense to stay put, conserving energy, if there have been hardly any fruits locally for the past two months, or to travel long distances to look elsewhere. Making the best choice may not always be a matter of life or death, but always making the better choice certainly means a longer life, with more offspring who also tend to make the right choices.

And this in turn leads to a possible explanation for other facets of orang-utans, especially their unusual society, which is less social than that of almost any other primate. Large groups of orang-utans may not be able to survive as groups in the long

Left *Fig fruits are probably not a highly nutritious food for orang-utans, but they are produced frequently in large amounts by many different fig species, and so represent an important staple for orang-utans in a variety of habitats.*

term if individual fruit sources are not big enough to provide meals for a group, if fighting over fruits ensues, and if a good deal of individual experience and intelligence is required to find fruits in difficult times. This means that, to be sure of obtaining enough food in good and bad times, orang-utans are usually better off alone.

Another aspect of orang-utans is that they breed very rarely, and have only one offspring at a time, which takes many years to be able to survive independently of its mother. That means, so the theory goes, that a mother looking after a baby cannot afford to have to share food with other adults, and that youngsters need years to develop the required experience and food-finding intelligence.

As with most humans, orang-utans cannot obtain all that they desire or need, and they cannot expect a delicious feast daily. If all orang-utans had to subsist on fruit all the time, the species would surely have become extinct long ago.

At some times, therefore, when there is little or no fruit available in the forest, orang-utans feed on a variety of other foods, notably leaves, bark (actually the thin juicy cambium on the inside), pith, flowers and insects. Several researchers have reported that wild orang-utans will, on occasions, feed on small vertebrate animals and birds' eggs. Wild orang-utans also occasionally eat soil, perhaps so that the clay particles can absorb toxins from the bitter plant foods.

Sumatran orang-utans eat more fruits overall than do Bornean orang-utans, and the former seem to enjoy a more consistent, nutritious diet. The combined results of the various studies of wild orang-utans indicate that the Sumatran orang-utan diet is always around 70 per cent fruit, with 15 per cent leaves,

2 per cent bark and the remainder a miscellany of insects and other foods. In contrast, Bornean orang-utans have about 60 per cent of their diet as fruits, but with figures varying between 15 and 98 per cent according to season. The percentage of leaves in the diet of the Bornean orang-utan fluctuates between 0 and 70 per cent, while that for bark fluctuates between 0 and 55 per cent. The percentage of these latter items is probably even higher, throughout most of the year, in very degraded habitats.

The general belief among orang-utan biologists is that this difference is linked to the presence of a greater abundance of orang-utan fruit-food-producing tree species in Sumatran forests, and to generally more fertile soils in Sumatra, which tends to result in more fruit production per hectare there than in Borneo. Sumatran orang-utans also tend to feed on large amounts of few fruit species, with pulpy fruit being most popular. In contrast, Bornean orang-utans feed on a wider variety of fruit species, taking rather smaller amounts of each, and making do with more unripe fruits, and seeds as well as fruit pulp. During many years of observation by several researchers, Sumatran orang-utans have been recorded to feed on about 200 species of fruits. In contrast, a study of just over one year, of orang-utans in primary dipterocarp forest at Ulu Segama, Sabah, by pioneer researcher John MacKinnon, ended with a list of just over 100 food plants, including leaves, bark and other items. More recent observations of orang-utans in very damaged habitats in lower Kinabatangan, Sabah, by the Kinabatangan Orang-utan Conservation Project, have yielded a list of well over 300 species of plant consumed by orang-utans.

Left At times and in circumstances when there is very little or no fruit in the forest (as in this damaged peat swamp forest) Bornean orang-utans may switch to feeding primarily on leaves, bark and insects.

Opposite An orang-utan in Central Kalimantan feeding on the ripe fruit of Durio kutejensis, a durian species common in the non-swampy lowlands of southern Borneo.

In certain places and at certain times, a particular food type can feature prominently in the orang-utan's diet. Large strangler figs (banyans), those massive plants with thick aerial roots that sometimes reclaim ancient buildings, and that individually produce hundreds of thousands of palatable soft fruits at odd intervals, provide an important food source for orang-utans in some areas, but not in others. In the Suaq Balimbing forests of Sumatra, for example, strangler figs are very rare, and instead a gregarious peat swamp tree, *Tetramerista glabra*, known in the region as punah, provides the single main food source for orang-utans over large areas. In some areas, especially in hills, hard acorns (genus *Lithocarpus*) and chestnuts (*Castanopsis*) provide large quantities of nutritious nuts 'out-of-season'. In the swamp forests of lower Kinabatangan, Sabah, leaves of abundant climbing plants provide a significant food source year-round, in a habitat that has been transformed over recent decades into extensive secondary growth. In some parts of Borneo, a common pioneer tree that dominates re-growth forest after logging is *Anthocephalus chinensis*, variously known in the region as laran, jabon or kelampayan, and a likely significant timber tree of the future. Where this tree is abundant, orang-utans can often be seen feeding on the fruits (which provide very little nutrition) during the rainy time of the year, and on the bark at other times. Bark, however, has almost no nutritive value other than the small

amounts of juice that can be sucked from the tough fibre. We should not be too alarmed if wild orang-utans are eating a lot of leaves over a long period in damaged forests. But bark is a real famine food. The health and survival of orang-utan populations may be at serious risk if they have to rely on bark as a major food source over long periods.

Cheryl Knott of Harvard University and colleagues observed over 60 wild orang-utans during nearly 5,000 hours in one year in Gunung Palung National Park, West Kalimantan. During that period, there was a major peak of fruiting in the forest, followed by a trough with very little fruit. During the month of highest fruit availability (January 1995), all the orang-utans seemed to eat only fruits. In contrast, during May 1995, only about one-fifth of their diet was fruit, while the rest was mainly bark, leaves and pith. Nutritional analyses of 78 of the foods most commonly eaten during the study showed that mast foods (that is, species of fruits produced very abundantly) were significantly higher in caloric content than were other foods. During the month of highest fruit consumption, males consumed an average of about 8,400 kilocalories (kcal) daily, while females consumed 7,400. During the month of lowest fruit consumption, males consumed an average of 3,800 kcal a day, and females only 1,800. The orang-utans had to travel further to find food when fruit was scarcest.

Opposite *Male orang-utan feeding on termites in a* Pterospermum stapfianum *tree, Kinabatangan, Sabah.*

Left *Female orang-utan feeding on tree bark, a famine food for Bornean orang-utans, that allows them to survive during periods when there is no fruit and few young leaves available in the forest.*

A number of laboratory-based studies provide intriguing additional clues about the orang-utan diet, as well as its relationship to anatomy. For a start, Cheryl Knott added value to her field observations by collecting 257 urine samples from the orang-utans that she observed, taking the samples to a laboratory, and then measuring ketones in the samples. Ketones are present when body fats are broken down to produce carbohydrates, and the body breaks down the fats only when there is insufficient food intake to supply those carbohydrates directly. Ketones were found in the urine samples only during the period with least fruits available and eaten, providing a strong indication that the orang-utans were indeed stressed by inadequate food during that period. (In addition to that, Knott measured levels of oestrone conjugates in the urine samples from female orang-utans. Oestrone conjugates are an indicator of the production of eggs in the ovary. She found that all orang-

utan matings observed during the study period occurred during the time of high food availability, which coincided with high average oestrone levels.)

Researchers interested in tooth abnormalities in primates have noted that living and extinct great apes are prone to an enamel imperfection called linear enamel hypoplasia. One study showed that wild caught orang-utans and chimpanzees have much higher rates of tooth enamel imperfection than other primates from the same area. A part of the study included a collection of 97 primate skulls obtained from the lower Kinabatangan, Sabah, in 1937. Another study suggested that this feature is a manifestation of temporary stress occurring at roughly six-monthly or yearly intervals from age 2.5 years onwards in wild orang-utans. The studies suggested a link to periodic weather and food stress, or perhaps also disease linked to changes in weather and diet.

Andrea Taylor of Duke University Medical Center noted that orang-utans exhibit the greatest variation in skull measurement among living great apes. Comparing jawbones of Sumatran and Bornean orang-utans, she found that the shape and size of the Bornean ones seemed better adapted to a tough, fibrous diet. This also suggests that the differences between Sumatran and Bornean forests in terms of orang-utan foods are real and have been constant for a very long time.

From the various studies outlined above, we might be tempted to conclude that Bornean orang-utans have been stuck for a long time with a habitat less suitable than that enjoyed by their Sumatran cousins, and that all wild orang-utans, especially the Bornean, live a feast and famine lifestyle, gorging on fruits whenever possible so as to tide them over bad times of near-starvation. An ingenious but simple

experiment performed with the help of three Sumatran orang-utans (an adult male, an adult female and a juvenile) by a group of scientists in Missouri, USA, showed that this is not quite the case. The orang-utans were given five diets with different proportions of fibre. Faeces produced after each type of diet were analysed to determine the levels of volatile fatty acids, substances which are absorbed and utilized by the body, and an indicator of how much the fibre in the diet had been broken down by fermentation in the hind gut. Surprisingly, the amounts of fatty acids were always similar, whatever the diet, showing that the orang-utan's digestive tract can digest fibre and use it for energy.

Combining all the above information, we can make a prediction that field workers concerned with the conservation of orang-utans may wish to address. It is that the Sumatran

orang-utan is more likely than the Bornean orang-utan to be adversely affected by intensive logging of forest habitat, because the latter seems better adapted to a varied diet of leaves and hard fruits, which can be found readily in damaged forests. Another pair of observations for people concerned with orang-utan survival in the extensive degraded habitats that now occur in many parts of the orang-utan's distribution are, first, that sustained bark-eating is an indicator of serious trouble; and second, that leaves of pioneer plants represent a better kind of food for orang-utans, especially the Bornean, than was previously believed, because these leaves contain protein and a moderate fibre content that can be at least partly digested. Forest rehabilitation efforts should ensure the presence and regeneration not only of trees, but of any plants whose leaves are often consumed by orang-utans.

Opposite (left) *The form and dimensions of the teeth and jaw bones of the Bornean orang-utan (shown here) suggest that this species is better adapted than the Sumatran orang-utan to a tough, fibrous diet.*

Opposite (right) *Experiments involving the food and faeces of three Sumatran orang-utans in a zoo indicated that the species can digest dietary fibre.*

Above (left) *Sumatran orang-utans may be more sensitive to forest degradation if fruit supplies are reduced, but they may be able to switch to a more leafy diet than previously believed.*

Above (right) *Bornean orang-utans seem better able to cope with poor-quality foods resulting from forest degradation.*

ORANG-UTAN SOCIETY AND REPRODUCTION

Orang-utans are very intelligent, have a strong bond with their mother, and are rarely observed to be aggressive to one another (the exception being adult males competing over a receptive female). One would imagine, therefore, that they spend much time together, interacting in groups as monkeys and the African apes do, but perhaps with more grace. In fact, orang-utans are famed among zoologists as being the least sociable of all primates, among the least sexually active, and with the slowest birth rate, not only among primates but probably among all mammals.

The basic features of wild orang-utan society and reproduction were observed and recorded by the two pioneers of wild orang-utan observation, John MacKinnon (who observed Bornean orang-utans in Ulu Segama, Sabah in 1968–71) and Herman Rijksen (who observed Sumatran orang-utans in the early 1970s). Many other researchers have added to our knowledge since then, but the basic picture remains the same. Orang-utans do not live in groups of more than two individuals. If more than one orang-utan can be seen in one place at the same time in the wild, the explanation will normally be that they are a female with her offspring, individuals who happen to be feeding in or near the same tree, or a mature male and female in a temporary mating

relationship. A mature female orang-utan will normally be accompanied by a single offspring, as only one baby is born at a time. It remains with its mother at all times, either clinging to her body or close to her, until it is about two-and-a-half years old. Indeed, usually the best way to determine whether an orang-utan seen in the wild is an adult female is to look for an infant or immature orang-utan. If the orang-utan is not large, and if there is no obvious infant on her body, or a juvenile orang-utan within ten metres or so, the chances are high that the orang-utan is an adolescent, of either sex, in the process of leaving its home area.

Orang-utan pregnancy lasts for about 265 days, and the newborn infant weighs about 1.5 to 2 kilograms. Almost always there is a single infant but zoo records show that about one per cent of orang-utan births are twins. The mother is very protective of her new baby and, whether in the wild or in captivity, shields it from the view of anyone trying to watch. In the wild, orang-utan babies appear to grow more slowly than human infants, while bottle-fed orang-utan babies in captivity may grow much more quickly.

Above *The only lasting social bond among wild orang-utans is that between a mother and her infant, until the latter moves away from her at age around six or seven years to begin an independent life.*

The infant orang-utan

During its first two years, the wild orang-utan infant depends entirely on its mother for food. The infant drinks mother's milk, and occasionally takes small pieces of food from her mouth. A mother orang-utan always carries her baby when travelling through the forest. The infant instinctively clings to her fur, but she gives it support when necessary. Usually, the infant is positioned halfway between the mother's chest and side, with one hand gripping her upper back, the other hand gripping her chest, one foot gripping the lower back, and the other foot her belly. This allows the mother to swing through the trees without harming the infant, while the infant's mouth remains close to the comfort of her breast and heartbeat. The infant always sleeps with her in the same nest. The infant orang-utan's large, dark eyes, with long, sticking-out hair around the face, give it a tragicomic air. Before they are two years old, orang-utans start playing in their arboreal surroundings, sometimes with their mother, sometimes swinging on branches or manipulating twigs, leaves or bark. Later, the youngsters start making play nests and hats out of leaves. By the age of two-and-a-half, an orang-utan still spends most of its time with its mother, but starts to make short exploratory trips in tree tops on its own, usually within her view.

Like human beings, infant orang-utans develop a total of 20 milk teeth, starting after the age of about six months. Orang-utans grow their teeth much more rapidly than humans, however, with the full set of milk teeth complete by the age of one year, and most of the permanent teeth replacing them during the fourth and fifth year. Also like humans, the permanent set consists of 32 teeth.

The young orang-utan is not fully weaned from milk until it is about three-and-a-half years old. By the time it is six to eight years old, the young orang-utan will usually find that its mother gives birth to a new infant, presenting it not only with a small sibling to play with, but also the need suddenly to fend very much for itself. Because of the long birth interval, young orang-utans have little chance of making friends. But if another mother and young orang-utan come along, perhaps encountered in the same fruiting tree, youngsters will play together while the two mothers will virtually ignore each other.

Above *An adolescent orang-utan approaching maturity and now independent of its mother.*

Opposite *A mother and new-born infant.*

Adolescence

Orang-utans are regarded as adolescent – the equivalent of human teenagers – by the age of about six or seven years. Prior to this time, youngsters learn slowly, by copying their mother and by trial and error, how to find their own food, build their own nest, and generally how to move around and live in the forest. Adolescent orang-utans have some contact with their mother, but they possess all the skills necessary to survive alone, and they spend increasingly longer periods away from her. Female orang-utans reach sexual maturity at about seven or eight years of age, although not yet fully grown, and they will then gradually leave the mother to lead a more solitary existence. Field studies suggest that Sumatran orang-utans develop somewhat more slowly than Bornean orang-utans. It is probable that most Bornean orang-utans give birth to their first child when eight to ten years of age, but long-term studies by many researchers, at Ketambe, Sumatra, suggest that Sumatran females may be as old as 15 years before they first give birth. The same studies, and other studies in the Suaq Balimbing swamp forests of Sumatra, suggested that the average interval between births was eight to nine years. A wild female orang-utan may deliver and rear only three to five offspring during her lifetime. Male orang-utans start to leave their mother at a similar age to females, but will not be sexually mature until they are 13 to 15 years old, and may not have the chance to mate until well after that age.

A solitary life

Orang-utans are quiet animals which move, make noise and communicate with one another only when it is really necessary. Despite their solitary nature, orang-utans do not occupy individual territories but instead live in overlapping ranges within a particular region of forest. For example, 16 different adult females, 9 adult males and 15 sub-adult males were all seen within a four-hectare intensive study area (200 metres by 200 metres) in Sumatra. Thus, orang-utans do see one another during their travels, and in areas where there is a high population density of the species, only rarely will a day pass without a glimpse of at least one relative or acquaintance. Most often, orang-utans meet when coming to feed in the same tree, and on such occasions one of the most remarkable things is the lack of either greetings or of squabbles over food. The only exceptions appear to occur when an adult male meets another male orang-utan who is unknown or unrelated to him. Then, short but aggressive chases may occur. Young mature males may voluntarily leave a fruiting tree to avoid trouble with an elder

male, but this is often done with barely a glance between the two. Generally, orang-utans which know one another feed peacefully together. It is believed that orang-utans maintain a rough idea of what other members of their species are up to in the surrounding forest, through occasional calls and signs of fresh nests, as well as brief encounters on their day-to-day travels.

In Sumatra, and probably also in Borneo, females tend to cluster in the same general area for much of the time, even though they cannot see one another. There is also a tendency for females to give birth during the same periods, a feature perhaps linked to times of high fruit abundance. Orang-utans of all age groups seem to cluster together much more in Sumatra than they do in Borneo. It was speculated by early researchers that, by sticking together, the former may be safer from tigers, which are present in Sumatra, but absent from Borneo. But, since tigers hunt on the ground and there have never been any records of tigers killing wild orang-utans, this idea now seems highly unlikely.

For most of the time, the adult male orang-utan lives alone, while the adult female is caring for one youngster. From time to time, when the youngster is more than four years old and able to travel and feed by itself, a flanged male and a female will meet and, after some courtship, they remain together for a temporary period, during which mating occurs. Such 'consortships', as they are known, last from a few days to a few weeks in Borneo, and up to several months in Sumatra. It seems female orang-utans actually seek older, fully mature males to father their children. These are the large males that give occasional long, loud calls, usually during the morning but sometimes at night. The loud calls seem to serve simultaneously the functions of attracting receptive females and of warning other large males to stay away, and perhaps generally informing all relatives of the location of the 'old man'.

The only major departure from the prevailing quietness and gentleness of wild orang-utan society involves non-flanged males during their time between adolescence and full maturity, a period exceeding five years for most individuals. From as young as eight years, before they are sexually mature, male orang-utans may try to force themselves on sexually receptive females, including those carrying youngsters. Females usually try to resist, but the strength of the male often prevails.

The ability to obtain orang-utan DNA from faeces and to identify individual orang-utans from that DNA, coupled with ongoing field studies observing orang-utans in both Sumatra and Borneo, have allowed researchers to obtain better insights into

how wild orang-utans are related. There have been several surprises. One is that a high proportion of orang-utans – perhaps about half – appear to be sired by non-flanged males. So, while females clearly prefer to mate with flanged males, and individual flanged males sire more offspring than individual non-flanged males, the 'strategy' of non-flanged males to spend much of their time seeking and forcing themselves on females works for those non-flanged males. It is better for them to take that approach than to wait and hope to topple the dominance of the few, old flanged males. That implies that unflanged males will tend to move great distances, seeking as many females as possible, and perhaps developing flanges along the way as they extend into forest where there are no other males. That idea fits with data from Sumatra, but less so from lower Kinabatangan, Sabah, where males in one area are as closely related to one another as are females in the same area. Perhaps that is because the habitat in Kinabatangan is surrounded by oil palm plantations, so males are no longer able to move far from their birth place. It seems also that different males may adopt different approaches, perhaps based partly on their

genes and partly on local circumstances. If a male decides to wander far, opportunistically mating on the way, he may end up with big flanges but no females. He will then have to spend his life out of the breeding population and childless. Or he may instead return to the thick of a breeding population, and fight resident flanged males in order to replace them. All these possibilities help to account for another observation – that there is a male-biased sex ratio among orang-utans at birth, but that this gives way to a female-biased sex ratio among adults. Some males leave and never return to breeding populations, while others return, occasionally to kill or be killed.

Opposite *A distinctive feature of wild orang-utans is that they spend much of their life alone.*

Above *Female orang-utans prefer to mate with large, flanged males such as this, but up to half of all wild orang-utans may be sired by unflanged males.*

A DAY IN THE LIFE OF AN ORANG-UTAN

For an orang-utan, the day typically begins at dawn. Colours gradually become visible after the blackness of the night, and outlines of leaves, branches and trees take shape. The hour before dawn is the coldest time of the day, and mist still hangs in the forest canopy. Soon, the sun's rays will vaporize it away. Birds, insects and gibbons, taking the cue of the slowly penetrating light, are now bursting into song. On dull, rainy days the orang-utan may stay in its nest for an hour or two, to avoid getting drenched in cold water while beginning the day's travel. During long rainy periods, there is very little fruit in the forest and little incentive to make an early start. Nothing really dries out, and slippery surfaces everywhere bring to orang-utans a greater risk of a bad fall from the forest canopy.

Swinging upwards out of the nest, using both hands and feet, the orang-utan will often seek a firm branch for support, before urinating and defecating. Like other mammals, orang-utans will not normally soil their sleeping place. The orang-utan now swings and climbs with care towards the nearest fruiting tree. It moves quickly and expertly, but never in a rush. Its bulk and lack of tail for balance make jumping through the canopy too risky. Entering the fruiting tree, it may find other animals, such as gibbons or hornbills, have arrived even earlier, but this is no problem. Rainforest animals in the same tree generally avoid each other. And the orang-utan will find plenty of under-ripe fruits, ignored by the more selective birds and gibbons. After an hour or so of feasting, the orang-utan may sit in the tree for a rest, and then move off at a leisurely pace to seek different food sources. By now, it is mid-morning and becoming hotter. The orang-utan may start to seek a cooler part of the forest in which to have a siesta through the middle of the day. On the way, it may make a detour to pick a few cryptic fruits and young leaves from lianas, and to check on the progress of trees which will bear a fruit crop next month.

Since the orang-utan's bulk makes travel difficult in many parts of the tree canopy, it often takes short rests. From time to time, the travelling orang-utan may cross gaps in the forest canopy by rocking small trees or lianas to and fro until it can reach the next tree with an outstretched hand. Occasionally, but especially in damaged forests, orang-utans have to descend to the ground to cross a gap, but they do not feel comfortable doing this, just as humans do not feel safe crossing a rainforest stream on a thin log. When passing a tree with a small

depression in the fork of a branch, an orang-utan may take the opportunity to scoop out some trapped rainwater to drink. For its siesta, it may choose a ridge top that catches the breeze, or a thicket of lianas beneath the crown of a tall tree. Here it may stay for perhaps three hours or so, stirring occasionally to scratch or stretch its body. By mid-afternoon, the orang-utan may seek a little animal protein, starting a meandering stroll through the forest canopy, choosing the easiest branches and hanging liana stems to swing between awkward gaps. The orang-utan may find the cardboard-like nest of some tree ants, pull off a chunk and hold it to its lips, allowing a steady stream of ants to enter the mouth. Strangely, the orang-utan ignores bites to its lips, but quickly brushes off any stray ants that get into the hair. While approaching the ant tree, the orang-utan may have noted a noisy commotion of green pigeons in the forest canopy on the other side of the valley – this normally means the presence of a strangler fig plant with fruit. The orang-utan crosses a stream by a series of small overhanging trees, dangling over the water for a few minutes to take the coolness that it offers. The figs will be its last meal of the day, but there is no reason to rush, because the orang-utan can spend the night nearby, and take advantage of the fig plant for many days to come, as a main food source. During today's example, the orang-utan travelled, zigzag fashion, for about half a kilometre. One day's travel for most orang-utans, on most days, will vary between 300 to 1,000 metres. When moving from one general location to another, it will travel a longer distance and in a straighter line.

The sun starts to sink below the furthest tree tops and it will soon be difficult to see details in the forest. By this time, the orang-utan will normally make a nest in which to spend the night. The orang-utan descends from the large strangler fig canopy, grabbing the last few fruits as it goes. It has already selected a suitable tree nearby, 20 metres high, with small, soft leaves and several stout, forked branches. To make the nest, it sits in a branch fork and expertly breaks off surrounding smaller branches and twigs inwards to form a springy platform.

Opposite *Orang-utans travel through the forest canopy by a variety of methods, including swinging between lianas, feeding on small food sources such as flowers and leaves as they go.*

Far left *Infant orang-utans experiment with all aspects of their environment, under the guidance of their mother.*

Middle and above *A common method of progression through the forest is by rocking and swinging between vertical supports such as small trees and lianas.*

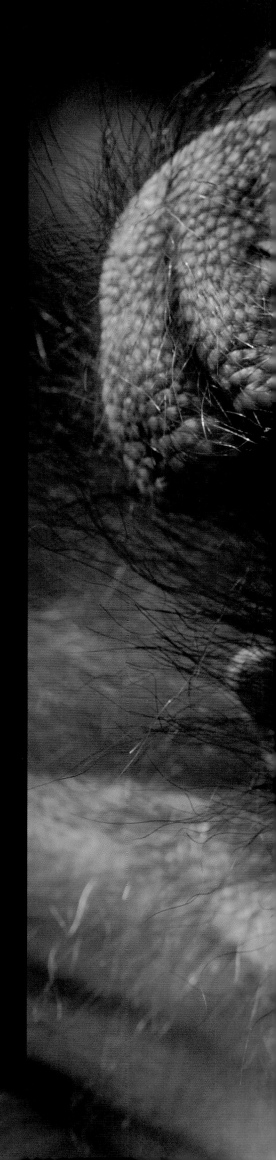

CHAPTER FIVE

ORANG-UTANS IN ZOOS

ORANG-UTANS ARE AMONG THE MOST POPULAR SPECIES IN ZOOS, PERHAPS PARTLY BECAUSE EVERYONE IS EQUAL IN THEIR PRESENCE. WHEN HRH PRINCE PHILIP, THE DUKE OF EDINBURGH, VISITED LONDON ZOO IN 1969, HE PEERED INTO THE ORANG-UTAN ENCLOSURE AND WAS SOAKED WITH A WELL-AIMED SPRAY OF URINE FROM A YOUNG MALE NAMED NAPOLEON. ALTHOUGH THERE ARE DEBATES ON THE ETHICS OF KEEPING APES IN CAPTIVITY, ZOOS HELP PROVIDE NEW INFORMATION ON THE SPECIES AND ARE NOT A DRAIN ON WILD POPULATIONS.

Until the early years of the 21st century, the majority of orang-utans kept in captivity did not survive for many years. Poor nutrition, harsh conditions and diseases took a heavy toll. In 1928, three orang-utans gave birth at Berlin, Nuremburg and Philadelphia zoos. These were the first zoo births, and official records show that between that time and December 2006, 1,644 orang-utans were born in zoos.

Nowadays, zoo orang-utans are managed as if all individuals are part of two captive populations, one Bornean and one Sumatran. This is done through regional captive breeding programmes managed by the American Zoo and Aquarium Association (AZA), the European Association of Zoos and Aquaria (EAZA) and the Australasian Regional Association of Zoological Parks and Aquaria (ARAZPA). Each association maintains an orang-utan studbook.

The World Zoo and Aquarium Conservation Strategy encompasses several areas in which zoos have conservation functions, including: maintenance of global populations so as to minimize genetic loss; the ambassadorial and educational roles of zoos; using captive populations to further biological knowledge of the species; and linking captive populations with conservation activities in the wild, such as re-introductions or fund-raising for wild conservation efforts.

Pregnant orang-utans in zoos draw special attention and care, as do mother and baby from birth onwards. Occasionally, zoo orang-utan births are done by caesarean section if there are complications. Professional gynaecologists and anaesthetists are normally brought in to perform the operation. In the case of one zoo orang-utan born in this way in 2005, the 15-year-old mother had never given birth before. In order to ensure that she would be able and willing to raise her first child, many zoo staff and volunteers took turns to help and teach the mother what she should do.

AN ETHICAL DEBATE

Watching orang-utans in a zoo may remind us of clowns, jokes or delinquent children, but the keeping of animals in captivity, especially intelligent mammals, is an issue that can raise polarizing debate, both emotional and rational. Some feel that it is wrong to keep intelligent animals in restricted spaces in unnatural

Previous pages *Orang-utans receive much love and attention in well-run zoos.*

Opposite *Most modern zoos adopt high standards and policies relating to animal acquisition, welfare and relations with humans.*

Left *Many people feel that it is wrong to keep intelligent animals in restricted spaces in unnatural conditions; cages such as this should be used for orang-utans only as a temporary measure.*

Below *Cages may be unsightly (although with good management fairly hygienic), but who is to say that an orang-utan cared for in captivity is necessarily worse off than a wild one starving in a forest in the middle of a drought?*

conditions. But there is a historical context to the story and a variety of mitigating factors associated with the presence of orang-utans in zoos. Perhaps the place to start is the fact that thousands of orang-utans have lost their original forest homes in recent decades, yet most reputable zoos have steadfastly refused to accept any of these displaced individuals. A few good zoos have accepted small numbers on assurance that those individuals were rescued from dire circumstances and have been provided in accordance with all relevant laws. There are several reasons for this tough line. One is that the exporting and importing of wild-caught orang-utans across international boundaries is virtually prohibited under the provisions of Appendix 1 of CITES (the Convention on International Trade in Endangered Species of Wild Fauna and Flora), to which almost all countries are now signatories. Between 1975 and 2003, 324 orang-utans were moved between countries with CITES permits, of which 249 had been born in captivity. The trend has been a clear decline in the numbers of orang-utans moved internationally with CITES permits. Another is that moving wild apes to any zoo is considered to be either plain wrong, or sending the wrong signals, by many wildlife conservation organizations. Thus, zoos are not a threat to wild orang-utans.

Of course, there are menageries, circuses and entertainment establishments in various parts of the world that have accepted orang-utans under dubious or illegal circumstances, and which keep those orang-utans in reprehensible conditions. But such cases are shunned by all well-known zoos, globally. The bad ones represent the bad apples that can affect any type of institution or profession.

Apart from the proprietors of the bad establishments, it is probably fair to say that most zoo keepers are extremely devoted and have great empathy and compassion for the orang-utans in their care. Who is to say that such attention in a confined space is worse for an intelligent being than being stuck, perhaps with a dying infant and suffering from malaria, in a forest in the middle of a drought with no food or water?

There are a number of thoughtful people who argue that humans have no right to imprison any great apes. If one adheres to this belief, it is difficult to know what action should be taken in the case of orang-utans in zoos. Extreme options – such as leaving the enclosure door open and stopping the provision of food – are impractical. But can and should zoo orang-utans be taken and released into a forest in Borneo or Sumatra? There are indeed dedicated persons who have devoted many years to training formerly captive orang-utans for release into natural forest, where the emphasis has been on young individuals displaced from their natural habitat. Adult orang-utans which have spent much or all their life in a zoo, and young ones born in a zoo, are a different matter. Unless enormous time, care, planning and attention to detail can be assured for each orang-utan involved, it would be wrong simply to put zoo orang-utans into tropical rainforests, of which they would have no relevant

experience and in which they would have little chance of survival. In fact, there are now hardly any young zoo orang-utans which were born in the wild. Of 321 orang-utans recorded in European zoos in December 2005, only 38 were wild-born, all of which were more than 20 years old except for six, which were obtained from inappropriate captivity in Taiwan. Similarly, of the 212 orang-utans recorded in North American zoos in December 2006, only 14 were wild-born, all of which were at least 40 years old. The last wild-born orang-utan to come into a North American zoo was a male named Russell, which was confiscated at an airport in 1980 by the US Fish & Wildlife Service.

A typical diet for zoo orang-utans consists of a mix of specially prepared nutritious pellets, along with fruits (such as apples, bananas, oranges and grapes) and vegetables (such as green leaves, carrots, beans and sweet potatoes). An adult orang-utan would normally be given between 1.5 and 3 kilograms of food daily. With the advent and development of improved conditions and health care, zoo orang-utans tend to be larger and heavier than their wild counterparts. Many zoo orang-utans are obese, a consequence of their daily high-quality food and insufficient physical activity.

On balance, the world population of orang-utans in zoos is in a better condition than ever before. Average age at death has

increased; of 908 orang-utans known to be held in zoos globally in 2006, 187 were over 30 years of age. Satisfactory breeding rates and successful rearing of infants to adulthood in zoos mean that there is little need to obtain more orang-utans from sources outside zoos, except perhaps for the purpose of boosting genetic diversity within the global zoo population. Successful production of offspring by female orang-utans above 30 years of age is now regarded as normal in zoos. However, the overall global zoo orang-utan population – at least according to zoos that provide information – does seem to be fairly stable. For example, the known orang-utan population in European zoos was 309 individuals in 1989, and 321 in 2005. Many zoos exchange orang-utans from time to time, in order to minimize risks from inbreeding. Apart from the naturally long inter-birth interval of orang-utans, there are other factors that limit growth in the overall zoo orang-utan population size. One is that most zoos have limited space and resources, and individual zoo managers may not wish to increase numbers. Another is that there were formerly significant numbers of hybrids between Bornean and Sumatran orang-utans. In 1991, 16 per cent of orang-utans in European zoos were hybrids. Following agreement to keep the two forms separate and not to allow further breeding of hybrids, this had been reduced to 8 per cent by 2004.

A problem experienced by zoos that maintain and breed several orang-utans is that of maturing males. After maturity, males may become aggressive towards one another and need to be kept separate. As is the case with wild orang-utans, zoo males reach full maturity much more slowly than females, and rarely acquire their full cheek flanges, beard, puckered forehead and throat pouch before the age of 15 years. Also, as in the wild, when two or more large males live in close proximity, only one becomes very large and develops these physical features.

There are no definite maximum age records for an orang-utan, as the oldest zoo animals tend to have come in at an uncertain age. In 2005, the oldest zoo orang-utans of known age were a Sumatran female of 53 years and a Bornean male of 45 years, but several individuals have died in recent years exceeding 53 years of age.

Opposite *Zoo orang-utans tend to be larger and heavier than their wild counterparts, a result of daily high-quality food and limited physical activity.*

Above *All good zoos can play a powerful indirect role in support of orang-utan conservation through education and influencing opinion towards conservation.*

Experienced zoo people know that old apes can become sad, even depressed, after losing a long-time mate , and where possible the staff will think of ways to help overcome this. One such case is that of a 45-year-old female orang-utan who lost her male companion. A cat was introduced to her, and the two soon took to playing and sleeping together. In the wild, it seems that after youngsters have left their mother, no two orang-utans will spend much time together for the remainder of their lives. Ecologists believe that this is because a solitary lifestyle represents the best hope for individual orang-utans in rainforests to always find enough food during lean periods, and avoid conflict with relatives over meagre food sources. But in zoos, where two or more orang-utans live in close proximity over decades, enjoying plenty of good food and company, close emotional relationships of a kind that we often think of as peculiar to humans, can form and last. These observations demonstrate how zoo-keepers can gain insights into the minds of orang-utans that might never be possible through formal research on wild orang-utans.

Above *Old apes in zoos can become sad, even depressed, after losing a long-time mate.*

Opposite *There is no substitute for seeing an orang-utan close up – zoos provide the only such opportunity for most people.*

ZOOS AND CONSERVATION

Where funds and other circumstances permit, zoos and zoological societies can make a contribution to conserving endangered wild animals, including orang-utans, and their habitat. Many have done so. One of the pioneers was Frankfurt Zoological Society, which took up such a role in 1970, concentrating on the rehabilitation of Sumatran orang-utans and, more recently, translocation of Sumatran orang-utans to forests where the species has become extinct. Many other zoos have played roles in supporting research, rehabilitation, translocation and other conservation projects related to wild orang-utans. Often the main problem for zoos in making such a commitment is that their own funds are severely limited, and their managers have to make choices from a whole host of deserving causes around the world. Decision-makers in zoos also say that it is difficult for them to know which of the various orang-utan field projects are most deserving, and which will turn out to have the greatest conservation impact in the long term.

All good zoos can play a powerful indirect role in support of orang-utan conservation. According to a study funded by the US National Science Foundation, involving 12 zoos and 1,400 visitors over a three-year period up to 2006, the act of going to accredited zoos and aquariums in North America has a measurable impact on the conservation attitudes and understanding of adult visitors. Despite the abundance of wildlife films now available, there is no substitute for seeing a live orang-utan close-up. The impact of such visits can help people – even if only a small percentage of zoo visitors – to become interested in orang-utan conservation and to form opinions on the variety of issues that will impinge on the link between consumers and nature conservation in the coming years. If just one in a million of the estimated hundreds of millions of annual zoo visitors globally is stirred by their visit to devote either their life, or a part of their wealth, to a particular conservation cause, and if some of those few decide upon orang-utans or South-east Asian rainforests, then the future of this species becomes more secure. In more general terms, public opinion on a range of issues can have implications for the long-term survival of wild breeding populations of orang-utans. Relevant issues include products made from tropical woods, bio-fuels, and whether or not it would be beneficial to have large-scale planting of trees in degraded parts of the tropics as a means to absorb carbon dioxide.

CHAPTER SIX

ORANG-UTAN CONSERVATION

MENTION THE NAME 'ORANG-UTAN' IN THE 21ST CENTURY, AND IMAGES OF THREATS TO THE SPECIES COME TO MIND: LOGGING, FOREST FIRES AND OIL PALM. BUT WHAT IS REALLY HAPPENING? FOR A CLEAR PICTURE, WE NEED TO UNDERSTAND THE HISTORICAL AND GEOGRAPHICAL CONTEXT, AND CONSIDER THE TWO ELEMENTS TO ORANG-UTAN CONSERVATION: SAVING HABITATS SUFFICIENT TO MAINTAIN BREEDING POPULATIONS, AND DEALING WITH INDIVIDUAL DISPLACED ORANG-UTANS.

REASONS WHY THE ORANG-UTAN IS ALWAYS A THREATENED SPECIES

We can guess that the massive decline in the distribution of orang-utans throughout South-east Asia over the past tens of thousands of years was a result of a combination of climate change and hunting by humans. It is worth remembering also that the distribution of orang-utans must have increased in some areas over the past few thousand years. Some of the peat swamp forests of Central Kalimantan that still support thousands of orang-utans are less than 5,000 years old. A review of three main features of the species' biology can show us why in more recent times the orang-utan has become a threatened species, why its existence has always been rather precarious, and why its future survival is now tied to the existence of large areas of secure forest in good condition.

Firstly, orang-utans can survive only in natural forest, and only some kinds of forest – mainly n lowlands and swamps – can support large breeding populations of orang-utans. That means that simply setting aside large areas of forest, for example for conservation of biodiversity in general or to protect water supplies, may not be enough to ensure survival of breeding populations. The forests that need to be targeted for retention are those where wild orang-utan populations are large and thriving. Of course, the lowlands and swamps are the areas that have long been favoured by humans for settlement and agriculture, even before the explosion in the human population that has occurred in recent decades.

Secondly, orang-utans require high-quality foods, notably fruits, yet, unlike birds and fruit bats, they cannot travel quickly over long distances to seek food. In regions where natural fruiting is infrequent and seasonal, such as tropical rainforests in extensive hill ranges, orang-utan populations may not be able to survive, even in the absence of hunting or disease. Orang-utans can and do eat young leaves and bark, but these are not their staple food requirements. It is unlikely that orang-utans will survive and maintain breeding populations without fruits in their diet.

Thirdly, orang-utans grow slowly and breed very slowly. Individual orang-utans produce fewer babies than almost any other species. This means that any kind of pressure that acts to increase the death rate, even to a small degree, can have an enormous impact on the survival of the population. Such pressures may be hunting, or diseases such as malaria, or natural events such as long droughts.

Looked at in this way, it is truly amazing that the species has survived over hundreds of thousands of years, in the face of the Ice Ages of the late Pleistocene, which must have seen some of the Borneo and Sumatra evergreen forests reduced in extent to patchy refuges surrounded by seasonal forests, grasslands, swamps, rivers and sea. And it is little short of miraculous that the orang-utan survived in the face of hunter gatherer humans, who spread into Borneo over 40,000 years ago, and many of whom over hundreds of generations must have regarded orang-utans as convenient packages of meat to be targeted because they could not fly or run away.

Previous pages *The major orang-utan conservation issue is to save habitats sufficient to maintain wild breeding populations.*

Opposite *Orang-utans require high-quality foods, notably fruits, but unlike birds and fruit bats, they cannot travel quickly over long distances.*

Above *Forests in the lowlands and swamps are the habitats that support the highest densities of orang-utans. Peat swamp forests in the southern half of Borneo now represent the most extensive remaining orang-utan habitats.*

Left *Over the past 40,000 years, orang-utans have experienced climate and sea level changes, shrinking and expanding of lowland and swamp forests, and human hunters – the pattern of the species distribution has been changing constantly. Recent land use changes should be seen in this historical context. There is evidence that mature male orang-utans survive and move periodically between forest patches in oil palm plantation landscapes such as this, whereas females and young do not.*

TRENDS UP TO THE 1980s

The establishment of non-governmental organizations to help conserve nature is a phenomenon which started during the 20th century. The pioneering Fauna Preservation Society, established in 1903, concentrated in its early decades only on African endangered species. The Malayan Nature Society was established in 1940, and The Sabah Society in 1960, but neither in their early years had any involvement with orang-utans. World Wildlife Fund (WWF), initially a single international entity based in Switzerland, was established in 1961 because, to quote its first report, 'during the 1950s it had become increasingly evident that the impact of human progress and development on the natural world had produced what amounted to a state of emergency for wildlife' and 'at the same time it had also become clear that the work of conserving nature was constantly hampered for lack of money'.

There was immediate concern over the orang-utan, as Asia's only great ape, which had never been studied in the wild. WWF's project number 21 involved supporting Barbara Harrisson (wife of Sarawak Museum director, Tom Harrisson) in 1962 for 'a survey of the reasons for the alarming decline in the numbers of Orang-utan, an investigation of the illegal traffic in live specimens, a study of methods of returning captured animals to the wild in the Bako National Park (in Sarawak), and general recommendations for saving these apes from extinction'. At the

same time, project numbers 14 and 35 involved a mission by one Oliver Milton to ascertain the status of rhinos and orang-utans in Sumatra. It was known in the early 1960s that the Sumatran rhino remained in only very small numbers owing to hunting, even though at that time extensive forest still remained in Indonesia and Malaysia. It is not clear now why WWF feared that the orang-utan was suffering an alarming decline at that time. In 1965, Barbara Harrisson was sent to Sabah under project number 100 'to assist in the neighbouring country of Sabah with advice and co-operation in setting up a permanent sanctuary for the existing orang-utan population'. This latter venture turned out to be a key factor in providing support for the orang-utan rehabilitation centre being established at Sepilok. On the basis of those initial efforts, the first report of WWF for the period 1961–64 concluded that a global population of 5,000–6,000 wild orang-utans existed at that time, 'giving cause for very grave anxiety'.

We know now that the WWF estimate was a gross underestimate of actual orang-utan numbers in the 1960s, that there were large wild breeding populations in many regions where biologists had yet to visit, and that the illegal traffic in live orang-utans was in that context a rather minor threat to the species' survival.

Barbara Harrisson may not have been aware of the work of Dwight Davis, who collected orang-utan specimens for the

Left *In the 1960s, wildlife conservationists believed that populations of orang-utans were declining as a result of illegal traffic in live young animals.*

Opposite *In 1965, a Sabah Forestry Department/WWF project led to the establishment, by the Department's Game Branch, of the Sepilok orang-utan rehabilitation centre, aimed at training orphaned young orang-utans to return to the wild.*

Chicago Natural History Museum over a total period of eight months in 1950 and 1956 in forests south of Sandakan, Sabah, or of the 1961 comments of Mr Peter Burgess, then Chief Game Warden of British North Borneo, made in the 21st anniversary issue of the Malayan Nature Society Journal. Davis reported that the 'orang-utan is still fairly common, at least in areas near the coast, in eastern North Borneo', and that 'they seem to be confined to the primary forest', while Burgess stated that 'there is evidence from the number and distribution of the peculiar nesting platforms which this animal builds in trees and from visual records and captures, that the species is distributed throughout the Colony and that the stock is a considerable one'.

Commercial logging of large trees in the coastal lowland forests of Borneo and Sumatra had been going on since the early 20th century, with either manpower or simple motorized winches and narrow-gauge railways being used to haul large logs to rivers and coasts, where they could be towed and loaded onto boats. But the pace and environmental impact of that early logging was limited by several factors. Trees, typically exceeding a metre in diameter, had to be felled and cut into shorter logs by men with axes, and it might take more than a day to fell one big tree. The logs then had to be hauled through the forest with ropes by teams of men for further winching or loading onto a railway or into a river. The latter could be done only on flat ground or gentle downward slopes, so trees on hill terrain were safe from loggers. All of that changed in the 1950s, when motorized chainsaws and bulldozer-type tractors were

brought into Borneo. Suddenly, the rate at which a forest area could be harvested was no longer limited by how many trees the men could fell and cut with axes. And almost all terrain, whatever its slope or location, and mostly out of bounds before, could now be reached by bulldozers, which could not only make roads on hill slopes but also pull massive logs up and down any terrain.

None of those reporters on the status of orang-utans in the early 1960s imagined the massive four-decade wave of commercial logging that was soon to pass over the islands of Borneo and Sumatra. By the 1970s, millions of very large trees were being felled annually in the lowlands and coastal hills of Borneo and Sumatra, mainly for export as round logs to Japan. In the process, tens of millions of smaller trees were flattened annually as bulldozers made trails through the forest to the big trees, and as the big trees fell onto the surrounding small ones. By the 1980s, the logging companies had moved into the interior hill ranges of Borneo, and the volumes of timber cut annually peaked to levels that may yet be seen again from future timber plantations, but will never be seen again from tropical rainforests.

Biologists and conservationists became alarmed during the 1970s at the impact of all this logging on orang-utans. Several field studies done in the 1970s and early '80s suggested that the numbers of orang-utans declined after logging. But those studies were based mainly in primary forest, with only brief investigations in recently logged forest. Some primatologists just assumed that logging would be devastating to orang-utans.

TRENDS FROM THE 1980s TO NOW

More detailed studies during the 1980s, especially in Sabah, revealed a rather different picture. It became clear that orang-utans tend to move away from active logging and, where possible, stay in better quality forest, but the overall population density of orang-utans remained similar in extensive, once-logged forest to what had been estimated prior to logging.

Studies done by WWF-Malaysia with the Sabah Forestry Department from 1980 to 1986 suggest that there were probably about 20,000 orang-utans in Sabah at that time. Subsequent studies carried out by Kinabatangan Orang-utan Conservation Project (KOCP) with Sabah Wildlife Department in 2002–03 gave an estimate of about 11,000 orang-utans. These estimates suggest that there has been a decline in the Sabah orang-utan population of over 40 per cent in 20 years.

Hunting is a minor concern for orang-utans in Sabah and there is no evidence of any organized trade. Disease does not seem to be a significant factor. The main reason for the decline is conversion to plantations of forests in the eastern lowlands, the former strongholds of the species. Orang-utans can live and breed in timber production forests, and even in very damaged forests, but they cannot survive in any kind of plantation.

The conversion of forests to plantations in Sabah was based on government policy decisions made between 1976 and 1984. The idea was to move Sabah's economy away from being dominated by the exporting of unprocessed logs to a diverse economy, with agriculture playing a much larger role. All the lowland forests designated for agricultural development at that time had already been logged. During that time, the significance of lowland forests to orang-utans was not yet realized. The general view of primatologists was that logging was highly detrimental to orang-utans, and that efforts should be directed towards retaining the remaining primary forests of Sabah, and Borneo in general.

A paper published by Benoît Goossens and co-workers in 2006, based on analysis of DNA from 200 orang-utans in Kinabatangan, Sabah, indicated that this orang-utan population had undergone a bottleneck, or severe mortality of orang-utans, in recent centuries. It was suggested that the northern Borneo orang-utan population had declined by over 95 per cent in recent times as a result of human pressures. Whether or not such a drastic decline occurred, and whether it was over a small or large area, this study indicates how vulnerable the species can be to shocks such as forest loss, hunting, disease or famine.

Now, almost all remaining areas with orang-utans in Sabah have secure legal status in the sense that they are Forest Reserves, Parks or Wildlife Sanctuaries. The bulk of the orang-utan

Opposite *The pace and style of logging natural forests in Borneo and Sumatra changed in the 1950s with the introduction of bulldozers, which can build roads and pull massive logs on almost any terrain.*

Above *Orang-utans survive well in the very disturbed and fragmented floodplain forests of lower Kinabatangan, Sabah, but not in adjacent plantations.*

population lives not in the Parks or Sanctuaries, but in Commercial Forest Reserves allocated for wood production. These Forest Reserves have in most part been extensively and intensively logged. One (Deramakot) is certified as sustainably managed, under Forest Stewardship Council criteria, while the Sabah government aims to secure all other orang-utan habitat within Commercial Forest Reserves and to work on a long-term programme of forest rehabilitation jointly for orang-utan conservation and sustainable wood production. The issues associated with the 1,100 or so orang-utans living in the lower Kinabatangan area are outlined in Chapter 4.

The situation is Sarawak is much simpler. The remaining population of orang-utans, estimated at between 1,300 and 2,000 individuals, is almost all secure within two adjacent existing protected areas, Lanjak-Entimau Wildlife Sanctuary and Batang Ai National Park.

Trends in Indonesia are much more complex and difficult to address in favour of orang-utans. A 'population and habitat viability analysis' carried out by orang-utan experts in Indonesia in 1993 estimated that there were between 10,300 and 15,500 Bornean orang-utans and about 9,200 Sumatran orang-utans at that time. We now know that those estimates were too low for that time. But the executive summary from a report on a repeat of the whole population and habitat viability exercise, carried out by orang-utan specialists in January 2004, summarizes what happened. 'Serious downward trends in the integrity of Indonesia's forest estate occurred throughout the 1990s due to widespread logging and conversion to plantation

agriculture. Some protected areas were left relatively unscathed, while others suffered from devastating fires that resulted from unwise land use practices. Since the change of government in 1998, however, conservation in Indonesia has seen a virtual collapse, and deforestation has been enormous regardless of the legal status of the land. As a result, wild orang-utans are in steady decline due to logging, habitat conversion, fires and poaching.'

The 2004 analysis drew together better and newer information, and came up with the following results. In Borneo, *Pongo pygmaeus wurmbii*, of Central Kalimantan and the southern part of West Kalimantan, had the largest remaining population, with an estimated 32,800 individuals surviving at that time. However, this could arguably be the most threatened sub-species of Bornean orang-utan, because most of the remaining large populations occur in peat swamp forests, which will remain susceptible to drought and fire, even if they can be conserved free of logging and conversion to plantations. With an estimated total population in 2004 of between 6,650 and 8,435 individuals, *P. p. pygmaeus*, which occurs in southern Sarawak (1,300 to 2,000 individuals) and northern West Kalimantan (5,350 to 6,435 individuals), is certainly not secure, but has the advantage of being represented by one large,

Left *A large proportion of the northern sub-species of the Bornean orang-utan (*Pongo pygmaeus morio*) lives not in strictly protected areas – like this male at Sepilok – but in Commercial Forest Reserves.*

Above *Orang-utans can live and breed in wood production forests as long as there is always a diversity of tree species, and harvesting of logs is sustainable.*

Opposite *Numbers of the western sub-species of the Bornean orang-utan (*Pongo pygmaeus pygmaeus*) are lower than those of the other two sub-species.*

protected population across the international border In 2004, the Sabah estimate of about 11,000 individuals of *P. p. morio* was made, while another 4,335 individuals were believed to survive in several separate blocks of forest in East Kalimantan.

The 2004 habitat and population viability analysis suggested that about 7,500 Sumatran orang-utans remained, scattered among 13 separate forest blocks, and with only seven of those blocks containing more than 250 orang-utans. (This estimate excluded a new population of several tens of orang-utans in Bukit Tigapuluh National Park, consisting of rehabilitant orang-utans translocated into habitat from which they seem to have gone extinct during the 19th century.) Estimates of Sumatran orang-utan numbers made by other specialists at that time were more pessimistic. Whatever the exact numbers of orang-utans then and now, the Sumatran orang-utan does appear to be more seriously endangered than any of the Bornean sub-species, owing to clearly smaller numbers within a geographically small area, many threats to all the 13 populations, and possibly an inferior ability to adapt to forest degradation.

ILLEGAL CAPTURE AND TRADE

Indonesian law has provided protected status to orang-utans since 1925, while laws of Sarawak and Sabah did the same from 1957 and 1963 respectively. All these laws placed strong restrictions on humans capturing, killing, harming and keeping orang-utans. A 1990 Indonesian law further strengthened protection for orang-utans, setting maximum penalties for harming or capturing an orang-utan at a fine of 100 million rupiah and five years' imprisonment. Similar strengthening of state laws was implemented in Sabah in 1997 (Wildlife Conservation Enactment, with a 50,000-ringgit fine and five years' imprisonment), and Sarawak in 1998 (Wildlife Protection Ordinance, with a 30,000-ringgit fine and two years' imprisonment). The Sabah law also refers to the impact on orang-utans of conversion of forest to plantations, requiring that developers take steps to mitigate forest loss, including bearing the costs incurred if orang-utans are translocated to other sites.

Malaysia has been a signatory to CITES (the Convention on International Trade in Endangered Species of Wild Fauna and

Left *Orang-utans have been protected by law for many decades in Indonesia, Sarawak and Sabah. Updated laws in all three areas provide heavy penalties for capturing or harming this species.*

Opposite *The Sumatran orang-utan is the most endangered of the living forms, with small and mostly declining populations scattered among 13 forest blocks.*

Flora) since 1977, and Indonesia since 1978. The CITES listing of orang-utans under Appendix 1 means that the species may not be moved between countries except for conservation-related purposes. Both countries therefore are committed to preventing trade in orang-utans.

Strong laws have been insufficient in Indonesia to prevent capture of wild orang-utans and their export to other countries, mainly as exotic pets or for animal shows. To some extent, this situation has been a symptom of several trends within Kalimantan, and Indonesia in general. One is extensive and repeated logging, coupled with conversion of forests to plantations, which means that wild orang-utans have still been losing their natural habitat through the 1990s up to the present. Dry periods and fires, especially during 1997 and 1998 but in following years as well, see stressed and starving orang-utans coming out of the forest to seek food in gardens, plantations and on roadsides. The few orang-utan rescue and rehabilitation

centres in Indonesia cannot cope with the numbers of orang-utans involved. Added to all this is the fact that there are many people in Indonesia, both local residents and new economic migrants into Kalimantan and Sumatra, who have very limited sources of income, which is why so many are involved in illegal logging. A few entrepreneurial individuals see an alternative source, in the form of trading wildlife. Orang-utans are a highly favoured species for local wildlife traders – they are not dangerous, they eat simple food and they fetch high prices. One orang-utan can bring in at least the equivalent of one month's income from a normal labouring job. While some orang-utans end up as pets for wealthy urban people within Indonesia, many are exported to other countries.

It has been estimated that, between 1985 and 1990, about 1,000 baby orang-utans were smuggled to Taiwan from Kalimantan for sale as exotic pets. The reason for this surge was in large part the result of a popular Taiwanese television

ORANG-UTAN REHABILITATION AND TRANSLOCATION

Two ways to help save orang-utans are rehabilitation and translocation. Both of these mainly concern efforts to secure the welfare of individual orang-utans, and both are made necessary by loss of forest.

The term 'orang-utan rehabilitation' generally refers to active efforts by humans to take orang-utans (usually immature), already in some form of captive conditions, from undesirable or illegal circumstances, and to provide those orang-utans with care and skills that will allow them to be released into forest, and to survive unaided. In recent decades, the origin of the orang-utans involved in rehabilitation projects has usually been linked to conversion of forest to plantation or farms, or to fires in or around orang-utan habitat. In the 1960s, however, the idea of rehabilitation was simply to retrieve pet orang-utans held illegally and to get them back into the forest. The first rehabilitation project was established in Sarawak in 1961 by Barbara Harrisson, but this was subsequently phased out, and a long-term project initiated at Sepilok in Sabah in 1964. The latter was linked in part to Harrisson's early surveys of orang-utans for WWF, and in part to the establishment of a Game Branch in the Sabah Forestry Department in 1964 under Stanley de Silva. In 1971, the Indonesian government established a rehabilitation station at Ketambe in Sumatra, followed by another at Tanjung Puting in Central Kalimantan (1972), and another at Bohorok on the eastern side of the Gunung Leuser National Park, Sumatra (1973). Further stations were established at Semenggoh in Sarawak (1977) and Wanariset Samboja in East Kalimantan (1991).

All orang-utan rehabilitation centres have similar aims and methods, and the experience of all the centres is that each orang-utan has a different personality and background. Each orang-utan needs slightly different treatment, and widely differing amounts of time, to become competent in the forest.

programme that featured a live orang-utan as a pet and companion. In 1990, Taiwan passed a law making it illegal for orang-utans to be kept as pets, and most were confiscated and sent to rescue centres within Taiwan. But the combination of a ready supply of orang-utans in Kalimantan due to habitat loss, and many people without a source of income, encourages such trade to continue. For example, in an article in the Jakarta Post in March 2004, it was reported that 196 Kalimantan orang-utans had been found recently at a zoo in Thailand and 30 in the United Kingdom. At around the same time, it was reported that 23 orang-utans had been smuggled to Japan in hand luggage.

The government of Indonesia, together with TRAFFIC, the wildlife trade monitoring network, continues to call for greater awareness among the judiciary, enforcement agencies and general public to ensure that trade in endangered primates, particularly the orang-utan, is considered to be a serious crime. This message will help save the orang-utan, if not their habitat.

Opposite *The Indonesian government authorities, with support from non-governmental organizations, continue to face great challenges to protect orang-utans and their habitats – this is a police post at Tanjung Puting National Park.*

Left *Efforts to teach vital skills to infant orang-utans continue at the pioneering orang-utan rehabilitation centre at Sepilok, Sabah.*

after their release. Initial efforts were made in Sabah where, starting in 1993, orang-utans were moved by the Sabah Wildlife Department from areas under conversion to plantations into Tabin Wildlife Reserve, a 120,000-hectare forest that was known to have an unusually small wild orang-utan population.

In recent years, there have been greater efforts to combine both these processes, by choosing orang-utans that have been rehabilitated most successfully after some years of sustained effort, and releasing them into selected forests where there is a chance that they may contribute to maintaining a large, wild population with an enhanced genetic base. This has been done in Sabah, where orang-utans from Sepilok have been transferred to Tabin Wildlife Reserve, and a monitoring programme is underway. Now, Sepilok continues to be the centre for rehabilitation of immature orang-utans, while Tabin is the main area in Malaysia to which orang-utans are translocated. Work by Borneo Orang-utan Survival Foundation (BOS) has expanded to seek new sites in the northern parts of Central Kalimantan for translocation of rehabilitant orang-utans. Given that so many orang-utans have been displaced by forest loss, and that many lowland and swamp sites that once supported thriving orang-utan populations now seem doomed, these efforts merit strong support. More recently, a 14-year-old Sumatran orang-utan named Temara, born in Perth Zoo, Australia, was selected for gradual release into protected forest in Sumatra.

The term 'translocation' generally refers to active efforts by humans to catch wild orang-utans (usually mature or adolescent) living in sites where they have little or no hope of long-term survival, and to move and release them into secure forest areas where they will have a fair chance of surviving and breeding. Orang-utans may need to be translocated either because the extent of remaining forest is too small to sustain them, or because numbers are too low for long-term breeding, or the habitat in which they are living is scheduled to be converted to other land use. Although the orang-utans may receive some degree of screening and health care during the translocation process, the general idea is to minimize contact between orang-utan and humans. As long as skilled people are involved, and an established protocol of actions is implemented, the risks to orang-utans are quite small. The main issues of concern are choosing a suitable forest area for the translocation, and knowing what happens to the orang-utans

Left *The Nyaru-Menteng project in Central Kalimantan involves the rescue, rehabilitation and translocation of orang-utans displaced by forest loss throughout the southern parts of Kalimantan. Through his large size and experience, this male has learned to make the most of the variety of foods provided.*

Opposite (top) *Younger orang-utans at Nyaru-Menteng require daily supervision and training activities – one small benefit of their plight is that they get to socialize with their peers in ways impossible for wild orang-utans.*

Opposite (bottom) *Human baby-sitters get to know every orang-utan individually at Nyaru-Menteng.*

Above Older orang-utans at Nyaru-Menteng, largely able to look after themselves, live on forested islands; there are too few trees to provide adequate food, which is instead supplied daily in the form of fruits and vegetables by boat.

Left A side benefit of the Nyaru-Menteng project is that jobs and small business opportunities are provided to local people – such as the provision of daily fresh food for the orang-utans.

ORANG-UTAN REHABILITATION CENTRES

Sepilok (Sabah, Malaysia)

This was the first orang-utan rehabilitation centre, started in 1964 on the boundary of the 4,300-hectare Sepilok Forest Reserve. Global interest, local demand, easy accessibility, cheap air travel and the successful development of tourism in Malaysia have all helped Sepilok to become a park popular with local and international visitors alike, with a reception area, cafeteria, signed trails, and a Rainforest Discovery Centre about one kilometre away. Visitor fees provide a large proportion of running costs. The centre is run by the Sabah Wildlife Department and supported by the UK-based Sepilok Orang-utan Appeal.

Tanjung Puting (Central Kalimantan, Indonesia)

Camp Leakey, on the Sekonyer River within Tanjung Puting National Park (about 300,000 hectares of peat swamp, heath and lowland dipterocarp forest), was established in 1971 by Biruté Galdikas for a pioneering study of wild orang-utans. A rehabilitation centre was established at the same time by agreement with the local authorities; additional rehabilitation sites were established in the 1980s. A decision was made in 1991 to stop accepting new orang-utans for rehabilitation, owing to the existing high densities of wild and rehabilitant orang-utans. In 1995, new regulations prohibited the introduction of orang-utans into areas with existing wild orang-utan populations. Since 1999, over 140 orang-utans have been released at five sites in the Lamandau forests to the west of Tanjung Puting.

Bohorok (Sumatra, Indonesia)

Situated at Bukit Lawang, on the eastern side of the Leuser Ecosystem, a rehabilitation project was started in 1973 with assistance from Frankfurt Zoological Society and later WWF. The site is run by the Indonesian Nature Conservation Agency. From the 1980s the site was seriously affected by tourism and increasing human presence. Acceptance of new orang-utans stopped in 1996, but a staff presence helps to maintain the earlier rehabilitants and manage tourists. The site suffered a devastating flood in 2003. A feeding platform used by the semi-wild orang-utans now provides a focus for tourists, while tourism helps support the local economy. The site is supported by the Sumatran Orang-utan Society.

Semenggoh, also known as Semenggok or Semengoh (Sarawak, Malaysia)

This is a small rehabilitation centre, set up (1975) and run by the Sarawak Forestry Department at the 640-hectare Semenggok Forest Reserve, which also has botanical research plots.

Wanariset Samboja (East Kalimantan, Indonesia)

Situated at a forest research site, next to the road between Balikpapan and Samarinda, this centre was established in the 1980s by the Indonesian Ministry of Forestry, the Indonesian Association of Forest Concession Holders, and the Tropenbos Foundation of the Netherlands. An orang-utan rescue and rehabilitation project was initiated here in 1991, with an emphasis on medical screening and treatment, limitation of human contact, and release into other areas free of wild orang-utan populations. Releases are carried out at Sungai Wain protection forest (3,500 hectares of prime forest) about 25 kilometres away, and into the northern part of the Meratus Mountains range.

Nyaru Menteng (Central Kalimantan, Indonesia)

During the devastating El Niño fires of 1997–98, hundreds of infant and juvenile orang-utans were rescued from villages in East, Central and West Kalimantan. Their mothers had become easy prey, forced out of the burning forests into human settlements, seeking food. Villagers, hungry and sometimes starving as a result of the prolonged drought, found easy protein literally in their backyard. While mother and other older orang-utans were killed and eaten, the juveniles were held in cages for future sale. Rescue operations were established, and hundreds of orang-utans were rescued by the Wanariset Project (above). It became apparent that a new quarantine and reintroduction facility was needed to cope with the continuous flow of orang-utans rescued from many dispersed parts of Kalimantan. The Nyaru Menteng site was chosen to help fulfil that need.

Located near Palangkaraya, the capital city of Central Kalimantan, the 62-hectare, mainly peat swamp, forest site has been built up within an arboretum (a site where trees are cultivated) founded in 1988 by the Indonesian Ministry of Forestry regional office. The orang-utan rehabilitation project is funded and managed by the Borneo Orang-utan Survival Foundation (BOS) in co-operation with the natural resource conservation office of the Forestry Department. The project has a veterinary clinic, quarantine cages, a forest site for infants to learn how to climb, midway and socialization facilities, and also three forested islands on a nearby river for the placement of larger orang-utans. Nyaru Menteng staff believe that hundreds of orang-utans are held captive – by individuals, as prestige pets – in Indonesia, mainly from sources in Kalimantan.

Due to its central location in a region of oil palm expansion, inflammable peat and expanding human population, as well as to a dedicated staff and infrastructure, Nyaru Mentang is likely to remain the most active orang-utan rehabilitation centre.

Welcome to / Selamat Datang di

Camp Leakey

Established in 1971 by / Didirikan pada tahun 1971 oleh

Dr. Birute Mary Galdikas & Rod Brindamour

Funded by / Dana didukukung oleh

Orangutan Foundation International (OFI)

A non-governmental organization & / Sebuah
registered charity / Lembaga Swadaya Masyarakat

For more information please visit / Informasi lebih lanjut website kami
our website

www.orangutan.org

Orangutan Foundation   Pemkab KOBAR   Departemen Kehutanan

Mohon Jangan :
·Memberi Makan /Minum :
·Makan/Minum Di Dep
·Menyentuh Orangutan.
Peringatan: Jangan Mencebur Atau
Berenang Di Sungai Ada Buaya
Mohon Bawalah Sampah Anda
Bersama Anda.   Terima Kasih

Please Do Not :
·Feed The Orangutans ;
·Eat In Front Of The Orangutans ;
·Touch The Orangutans.
Warning: Do Not Swim In The River
There Are Lots Of Crocodiles
Please Take Your Litter With You
www.orangutan.Org   Thank You

Above The pioneering orang-utan rehabilitation site in Kalimantan was that at Camp Leakey, Tanjung Puting National Park. The site remains one of the most important orang-utan conservation areas in Borneo.

Left A small number of orang-utans can be seen at the Semenggoh wildlife centre in Sarawak.

Opposite The Bohorok orang-utan rehabilitation centre in Sumatra has undergone many changes and challenges since 1973; now, domestic and international tourism helps to provide local business opportunities and to secure the site.

CURRENT THREATS AND CONCERNS

Following criteria endorsed by the IUCN (The World Conservation Union), the orang-utan is classed under the Red List as an endangered species, based on the belief that there has been a reduction in overall numbers of wild orang-utans of at least 50 per cent over three orang-utan generations.

Much has been written in recent years on tropical deforestation and the impact on orang-utans of oil palm plantations. Expansion of these plantations has indeed affected orang-utans, but the story is not nearly as simple as implied in those reports that blame oil palm for the demise of the species (see Orang-utans and oil palm, page 137). Within Malaysia (Sabah and Sarawak), the species is now probably better classed as vulnerable, because the rate of orang-utan habitat loss has fallen off to a low level in recent years; there is almost no hunting of the species; and most of the remaining populations occur in forests to be retained as protected areas

or under natural forest management for timber production (see Differences between Sabah, Sarawak and Kalimantan, page 139)

The threats to orang-utans in Indonesia are more complex and severe (see Differences between Sabah, Sarawak and Kalimantan, page 139). We read of El Niño droughts, forest fires and thick smoke resulting from vast areas of burning and smouldering peat soils; these are indeed recurring threats. Less widely understood is that the exact and ultimate legal authority over forest land in Kalimantan may not always be clear cut. National, provincial and local government, as well as indigenous communities, poor settlers and oil palm companies, may all claim rights over a specific land area, yet under existing laws no single one of them can claim a final decision-making role. In such circumstances, the future of forest in any given area remains precarious, as no one has a strong case to defend it and some may have a strong case to convert the forest to alternative land use.

Left The survival of orang-utans in Indonesia has to be linked to provision of jobs and food for a very large and growing human population. Preserving all the forests will not achieve that. Choices will have to be made as to which forest areas to save and what types of alternative land use will provide the most livelihoods for the most people.

Below As species, humans and orang-utans can co-exist; the challenges for the coming years are to decide exactly which forest areas will be allocated for the latter, and how those forests might be managed to provide economic, social or environmental benefits.

The main underlying issue is that Indonesia has a very large and growing human population. Forest-dependent communities living in harmony with nature are a small and declining minority. The forest areas containing orang-utans in both Kalimantan and Sumatra represent ways for people with limited opportunities to make a living, most obviously by cutting and selling wood, and by settling and farming. More enterprising individuals may trade valuable forest products – which may include infant orang-utans. When large numbers of people have very little choice as to how to make an income and provide food for their families, the legal status of the land and forest – even if it were clear – cannot on its own be expected to represent an effective way to save orang-utan habitat. Big timber businessmen, whether legal or illegal, must take part of the blame for the degradation and destruction of orang-utan habitat in Indonesia. But even if the timber trade can be effectively curtailed and made sustainable according to international standards, Indonesia will still be faced with the need to provide alternative jobs and food for the many hundreds of thousands of families who now rely solely on wood sales and farming for their survival.

The additional threats of drought, forest fire, and capture and trade of orang-utans still exist for orang-utans, especially in parts of Kalimantan. But the key issues of security over land and forest, and of permanent jobs for poor rural people, outweigh those concerns. Indeed, deforestation, forest fires and trade in wildlife will never be solved unless the key issues of resource security and sustainable income are addressed.

Orang-utans (or their human supporters) and oil palm (or palm oil; the palm refers to the plant, the oil to the enormously flexible commodity that is yielded by the palm fruits) are often portrayed as implacable enemies without hope of compromise. But somehow, a balance will have to be struck between conserving a very special species and the economic imperatives of Indonesia and Malaysia.

For at least three reasons, attempts to halt expansion of oil palm plantations are very unlikely to succeed. Firstly, oil palms produce high yields of oil for little financial cost, as well as jobs for the rural poor in the tropics. Secondly, the world demand for vegetable oils will continue to grow indefinitely and this situation is never likely to change, owing to the wide variety of potential end uses of palm oil. Giving up eating ice-cream, chocolate or biscuits will have a negligible impact on efforts to save orang-utans. Thirdly, if oil palm planting were to be successfully banned tomorrow, land-owners would simply plant other economic crops instead.

So, not only oil palm but various other kinds of plantation are here to stay, and set to expand in area. Oil palm is certainly a crop of special concern, however, because it thrives particularly well in the lowlands of Borneo and Sumatra. The issue is really by how much the plantations expand, and where.

Four issues need to be understood. First, the planting of oil palm on a commercial scale in Borneo and Sumatra started in the 1960s, accelerated in the 1980s and took off in the early 1990s. By year 2000, oil palm plantations had been developed in roughly 2,150,000 hectares of the lowlands in Borneo, representing barely 3 per cent of Borneo's land area. This increased to 4 per cent of Borneo's land area during 2006. In other words, there has been a rapid expansion of oil palm plantations in some parts of Borneo, but there are still large areas under forest and other land use. Second, loss of forest to other land use has always been a feature of development in all nations. Even if oil palm did not exist, forest will still be converted to rice, rubber, timber plantations and a host of other food and industrial crops. The issue is more one of whether forest loss is primarily a symptom of poverty, or rather part of a development process that leads to a thriving diverse economy with a human population that doesn't need to chop down more forest for subsistence needs. Third, there is an enormous and growing world demand for vegetable oil, previously only for edible products and detergents, and now also for bio-fuel. Oil palms yields more oil per hectare per year than any other known edible oil-producing plant species, at low financial cost. Looked at globally, oil palm may make better sense than some other oil-producing crops. And fourth,

oil palm is one of the few major oil-producing crop plants that are perennial, with a replanting rotation period of about twenty years, unlike most other such crops, which have to be replanted annually. If oil palms can be bred not to grow too tall for harvesting and to continue to produce high fruit yields for many decades, then the rotation period can be made longer. Thus, oil palm plantations do not require annual exposure and disturbance of soil, and can be viewed in a global context as having some ecological benefits over annual crops.

Oil palm is a fairly robust crop that can grow on many soil types, as long as there is year-round sunshine and frequent rainfall or irrigation. It also needs semi-skilled labour, because there is little scope for mechanization of harvesting, and plantations produce fruits almost continuously. We need to think of ways to expand plantations into regions where there is poverty and no rainforest, or at least no forest of high conservation value, so that pressure for conversion of extensive forest in Malaysia and Indonesia is lessened. In Kalimantan, oil palm plantations and their spin-off industries could serve as a means to provide jobs for poor families who are currently dependent for their income on logging and subsistence farming. Thus, oil palm plantations can be seen as a part of the solution rather than simply one big problem.

Although the Malaysian Federal Land Development Authority (FELDA) was responsible for conversion of large tracts of lowland forests to plantations from the 1960s to 1980s, its aim to help eradicate rural poverty was in most ways successful. Indeed, families who were subsistence farmers in Peninsular Malaysia up to the 1960s and 1970s have been transformed, such that children of the first plantation settlers no longer accept life in a plantation, and many have moved on to new opportunities in service and manufacturing industries, far from their original community. By the 1980s, the growing problem for FELDA, as well as for commercial plantations, was that fewer Malaysians wanted to work in plantations, and labour had to be sourced from other countries, such as Indonesia. Another positive aspect of governmental and commercial large-scale plantation development has been that they are forced to contribute heavily to the cost of establishing the necessary infrastructure for access to new areas and new communities. Of course, such developments may not be seen as positive if one feels that rural people in Borneo and Sumatra are better off following their old traditions. The risk with that view is that the children of rural people may become progressively more separated from the opportunities enjoyed by the young generation in towns and other communities with good road access.

Above (left) The oil palm (Elaeis guineensis) is a palm tree from West Africa, which grows to 15 metres tall. It has a cluster of leaves interspersed with frequent flowers and fruits at the top.

Above (right) Each oil palm fruit is about the size of a plum, growing in massive bunches, each weighing 20–30 kilograms.

Left Palm oil is extracted from the fruits' outer fibrous pulp, while palm kernel oil comes from a hard seed in the middle.

Following pages Primary dipterocarp forest: a balance will have to be struck between retaining sufficient to sustain the natural biodiversity and allowing other land uses to sustain human livelihoods and national economies.

Malaysia

Malaysia still is a developing nation, where land use, economy and society inevitably change with time. However, the relative importance to the economy of the exploitation of natural resources is declining year by year. Malaysia's *Vision 2020* aims to make the country a fully-developed and economically diverse nation by the year 2020. Clear and secure property rights play a key role, both in Malaysia's economic success and in nature conservation.

Sabah

Based on soil surveys done by the British Overseas Development Authority in the early 1970s, it was found that nearly 30 per cent of forest land in Sabah, mainly the eastern lowlands, was suitable for plantations crops. Most of that land was allocated for development during the late 1970s, as part of a general strategy to diversify Sabah's economy away from reliance on timber. In 1984, the government of Sabah established the permanent forest estate, 51 per cent of Sabah's land area, consisting of land that would remain under forest cover, either for protection or for timber production in the long term. By that time, almost all forest outside that permanent forest estate had been logged. International experts were at that time unanimous in their opinion that orang-utans are very sensitive to logging and some argued that, if orang-utans are to survive, primary undisturbed forest must be maintained.

Repeated observations of orang-utans in logged forests by researchers in Sabah showed conclusively that the largest numbers of orang-utans there were in fact in logged forests. Following that, and despite great difficulties posed by the need for acquisition of land already granted to private owners and allocated for plantations, the government of Sabah eventually agreed in 1994 to provide a new sanctuary for orang-utans and other wildlife in the form of Kinabatangan Wildlife Sanctuary. Since then, the Sabah government has remained sensitive to the need to conserve orang-utan habitat in the face of pressure for economic development – which, as it turned out, has been linked with expansion of oil palm plantations as well as wood-based industries. Most significantly, the Sabah government has made a commitment to retain the habitat of the largest remaining Malaysian orang-utan population, in the 240,000-hectare Ulu Segama Malua Forest Reserves. An important factor in Sabah is that almost all land is either privately owned or already established as Forest Reserve, Park or Sanctuary. Now, over 95 per cent of Sabah's orang-utans live in the Forest Reserves, Parks and Sanctuary.

Sarawak

Orang-utans in Sarawak suffered from chronic hunting pressure over thousands of years, which evidently led to their extinction in many areas, coupled with forests on predominantly poor soils that most likely provide food only for sparse and scattered breeding populations. By a century ago, breeding populations of orang-utans were already confined to the south of Sarawak, scattered in peat swamps, lowlands and the hill forests bordering with Kalimantan. These days, almost all of Sarawak's orang-utans occur in Lanjak-Entimau Wildlife Sanctuary and the adjacent Batang Ai National Park. The Wildlife Conservation Society is assisting the Sarawak government in surveys to provide better estimates of the numbers of orang-utans, reckoned to be between 1,300 and 2,000 individuals. A small population of orang-utans survives in peat swamp forest at Ulu Sebuyau, not far from the areas where Wallace and Hornaday shot their museum specimens in the 19th century. Proposals have been made for a new protected area here.

It is not anticipated that any ongoing or future logging or plantation development will have any impact on Sarawak's orang-utans.

Kalimantan

Although the forests of Kalimantan are very similar to those of Sabah and Sarawak, socio-economic conditions are quite different. Kalimantan is seven times larger than Sabah, with a much less developed infrastructure. Unlike Malaysia, Kalimantan does not have the concept of Forest Reserves, and rights granted over land to individuals and companies tend to be less clear-cut than in Malaysia. Indonesia has a very large and growing human population with many unemployed people. Overall, this means that, while Kalimantan has much larger areas of orang-utan habitat than Malaysia, the human pressures on those habitats are much greater. It is difficult for governmental authorities and NGOs alike to decide whether to try to save all remaining orang-utan habitat in Kalimantan, or whether to focus effort on a few selected areas. One NGO, WWF, has chosen the latter course, with three big 'orang-utan landscapes' identified for attention, while other NGOs are focusing attention on other sites. Whatever happens, there will continue to be significant expansion of plantations – oil palm as well as rubber and other crops – in Kalimantan. It makes sense for conservationists to engage and debate actively with the oil palm sector, both within Indonesia and globally, to help ensure that future plantation expansion will avoid all the major orang-utan populations.

CHAPTER SEVEN

WHAT CAN BE DONE?

ORANG-UTANS AS A WILD SPECIES ARE IN
SERIOUS TROUBLE IN PARTS OF THEIR RANGE —
A RESULT OF HUMAN POPULATION GROWTH,
POVERTY, ECONOMIC GROWTH, PLANNED
AND ACCIDENTAL FOREST LOSS, EL NIÑO
DROUGHTS AND FIRES, LOGGING OF
FORESTS ON LANDS THAT LACK SECURE
TENURE OR MANAGEMENT PLANS, AND
PLANTATION EXPANSION. INDIVIDUALS AND
SECTORS LINKED TO ANY OF THE ABOVE
NEED TO REVIEW THEIR INTERESTS AND
CONTRIBUTE TOWARDS SOLUTIONS RATHER
THAN LAY BLAME.

A WAY FORWARD

The well-known trend, whereby orang-utan habitats have been degraded and lost through logging, re-logging, fire and conversion to agriculture, is alarming in an ecological sense. Yet this trend is even more alarming than may be imagined. Once former forests have been degraded, they have little or no economic value to anyone. Generally, only wildlife conservationists see much merit in saving them. The best way to make such areas economically valuable – whether to subsistence farmers, plantation companies or governments – would be to rehabilitate them, so that they can produce future wood harvests, on a sustainable basis, with an agreed annual allowable timber cut. This may not involve replanting with the big, long-lived trees that formed the first crop, but rather with short rotation trees that will allow a harvest every ten to twenty years.

There is little doubt that a conceptual marriage is now needed, between making degraded forests economically valuable again and the imperative to retain and restore orang-utan habitat. We know that orang-utans can live and breed happily in damaged forests – the problem is that they may not be able to survive in the very damaged forests. The likely way forward is to rehabilitate the degraded areas with a few species of fast-growing native trees, for future harvesting, coupled with planting of prime orang-utan

food trees on sites that are less optimal for wood production, such as riversides, steep sites and permanent swamps.

The State Government of Sabah announced in March 2006 that logging will be phased out in the 2,400-square-kilometre Ulu Segama Malua Forest Reserves, which surround the world-famous Danum Valley Conservation Area. These forests support the largest remaining Malaysian population of orang-utans, estimated at over 4,000 individuals. WWF-Malaysia believes that the joint priorities now are to decide exactly how to best rehabilitate this forest at reasonably low cost, and to help Sabah get new sources of funds to carry out the rehabilitation work over large areas. Such treatment is not needed for totally protected areas such as Danum Valley and Lanjak-Entimau, which are in good condition and enjoy secure legal status. But there is now a clear need to determine site-specific visionary plans, particularly in Indonesia, to help secure the future for those damaged forests that still sustain some of the larger orang-utan populations. There are very few options. Combining orang-utan conservation with a plan for long-term sustainable wood production seems one of the best. This is the priority for Ulu Segama Malua in Sabah. Meanwhile, WWF-Malaysia field staff are collecting information on the orang-utan population through ground and helicopter monitoring of orang-utan nests over the whole 2,400 square kilometres, plus regular observations of individual orang-utans.

HOW WOOD-BASED INDUSTRIES CAN HELP

People involved in any wood-based industry, in any capacity, can help in three main ways. They can refuse to buy products that come from clearance of forest areas that contain or contained orang-utans; they can support certification of wood from sustainably managed forests, either as producers or buyers; and they can contribute ideas, support, decision-making or funding towards rehabilitation of damaged forest that contains orang-utans.

All wood products, including paper, can be made from a range of woods that can be grown in a variety of the world's climatic zones. There is no need, and no excuse, to convert any orang-utan habitat to plantation, for pulp, chip or wood production.

Orang-utans can survive and breed happily in natural forests that are well managed for timber production. This has already been proved since the 1990s, in the 55,000-hectare Deramakot Forest Reserve in Sabah, certified under Forest Stewardship Council principles and criteria. The logs from this forest are auctioned, and tend to attract premium prices. Meanwhile, the forest sustains one of the highest recorded population densities of orang-utans in logged lowland dipterocarp forest. All remaining forests in Borneo and Sumatra which are not protected areas and which contain breeding

populations of orang-utans should, if they cannot be made into protected areas, be allocated for sustainable natural forest management. In order to provide global market access to the wood from these forests, they are best certified according to the standards of an internationally recognized forest certification scheme. Buyers of wood and wood products from Borneo and Sumatra should endeavour to satisfy themselves that the wood has not been sourced from orang-utan habitats, unless the wood comes from certified forests. This does not mean that anyone should stop buying wood products from Indonesia or Malaysia. It means that anyone in the wood-based industries anywhere in the world can play their part towards halting forest loss and bad logging practices within the range of orang-utans, by saying no to doubtful sources.

Previous pages *Extensive oil palm plantations have replaced much orang-utan habitat, but the oil palm sector must now become a part of the solution to conserving orang-utans by helping lift subsistence farmers and loggers out of poverty.*

Opposite *A conceptual marriage is needed between restoring degraded forests, to make them economically valuable as sustainable wood producers and absorbers of carbon dioxide, and allowing them to maintain orang-utan populations.*

Above *Forest Stewardship Council certified logs in Sabah, harvested according to a management plan that matches the annual wood extraction rate to the natural tree growth rate.*

Certification of forests and taking steps to prevent trade in illegal wood products both represent important means towards achieving retention and sustainable management of natural forests. However, in isolation, those efforts must be supported by formal allocation and retention of a well-managed permanent forest estate, without repeated allocation of forest land for other purposes. Each 'forest management unit' needs an approved forest management plan. For example, the Ulu Segama Malua initiative of the Sabah government, to retain the largest Malaysian orang-utan population under a sustainable forest management regime, with investment in rehabilitation of degraded sites, deserves full support.

Damaged forests containing orang-utan populations can and should be restored, in order to revive timber production potential as well as to improve the habitat for orang-utans and to restore biodiversity. Private land-owners and forest managers can both take this step. Active forest restoration is beneficial, not only for orang-utans and the restoration of timber production potential, but because it demonstrates to those who advocate plantations on degraded forest land that there are people who care about restoring damaged forests, and will not just let them languish, or be open to future unsustainable logging or burning or conversion to plantations.

Rehabilitating damaged tropical rainforests does not make much sense in simple financial terms, because plantations normally yield more product than natural forests. Even plantations of timber trees will yield much more wood per hectare per year than natural forest. The only real benefit of forest rehabilitation over plantations, if the sole aim is wood production, is that plantations suffer the risk of epidemic diseases, wiping out the genetically very narrow resource. With this scenario, the task at hand is to ensure that several different kinds of goods and services can be obtained from natural forests if they are to be rehabilitated, so that not just a single user has an interest in retaining the forests. This is not an impossible task. Historically, the wood-based industry in Indonesia and Malaysia has had a wood-mining mentality that has not invested in the future, but has preferred to liquidate forest assets to get as much wood as quickly as possible. But this is already changing.

Restoring orang-utan habitat does not mean large investments in locking up land, because orang-utans can survive in restored forests that are to be managed for timber production. The important point is that the areas with orang-utans are not converted to plantations of one species, but instead treated to produce a mix of native tree species. These can be fast-growing species, as long as at least some are orang-utan food plants. Steep sites and river banks can be restored with favourite orang-utan and wildlife food plants. Such sites should never be logged in any case, if there is a good sustainable forest management plan, while further away from rivers, trees can be cut and sold according to a scheme

Left Where might further expansion of oil palm plantations occur in ways that will no longer jeopardize the survival of orang-utans? The Roundtable on Sustainable Palm Oil provides a promising global approach, if all stakeholders collaborate with an open mind.

Opposite Maintaining or restoring tree cover on marginal sites within oil palm plantations could contribute to orang-utan conservation in some circumstances – plantation decision-makers, foresters and conservation organizations need to join forces.

that does not require simultaneous clear-felling of large blocks. In many sites, promotion of tree growth can be achieved by cutting weedy plants such as climbing bamboo, without the need for expensive growing and planting of tree seedlings.

For conservation biologists who advise land-owners and forest managers, the immediate need is to find ways to restore forest plant diversity and growth of trees at the lowest possible cost in money. In some cases, this might be done by having a mix of fast-growing tree species on good sites, while planting strangler figs, the orang-utan's most important food source when there is little other choice, on rocky and other marginal sites. Of course, there will be a gap of ten to twenty years between planting and fruiting of these figs plants.

At least several native tree species exist in Borneo and Sumatra, which can be harvested after only twelve to twenty years to yield wood of known high commercial value. Forest managers, large plantation land-owners and small-holders all need to be encouraged to plant such trees on an economic scale, as a hedge against drought, floods and economic fluctuations, both for wood production and to play a role in recreating 'corridors' for genetic exchange and wildlife. Once the decision is made to restore damaged forest, the way is open for various income-generating opportunities, including the incorporation of such high-value forest products as agarwood, as well as nature tourism.

HOW THE PALM OIL INDUSTRY CAN HELP

The important questions for all concerned parties to address are the following: where might further expansion of oil palm plantations occur, and by how much, on both local and global scales, in ways that will no longer jeopardize the survival of orang-utans and other threatened species? All players in the palm oil industry should bear in mind that there are justifiable aims to ensure conservation of threatened species, to retain natural forests for their other many goods and services, to prevent risky plantation expansion onto marginal and ecologically-sensitive soils, to help diversify local and national economies, and to provide job opportunities in the poorest tropical regions.

There are several approaches that can be taken by oil palm interests. All have already been initiated and merit support. First, making the best use of what is already in place: improvement of yields per hectare in existing plantations, in reducing unnecessary loss of fruits between palm and mill, and in oil extraction ratios, are already strongly endorsed but need constant efforts for improvement. Second, further forest conversion must now avoid completely all sites where there are known breeding populations of orang-utans, whether in government-owned forests or private land. Third, further expansion of oil palm plantations should be targeted outside Malaysia and Indonesia, aiming at areas with

extensive non-forest or deforested land, and poor human populations with few economic opportunities. Oil palm can grow and produce oil on most substrates, throughout the tropics, as long as it is provided with the key inputs of water, fertilizer and labour.

Palm oil-producing countries and buyers alike can insist that plantation companies implement some form of robust environmental management systems, such as ISO 14000, which may be audited by credible third parties. Another channel by which the palm oil industry can help to reduce its impact on orang-utans is to promote interest to support the Roundtable on Sustainable Palm Oil. Members of this international, multi-stakeholder organization pledge not to be involved in conversion to oil palm plantations of primary and high conservation-value forest after November 2005 or – if they are buyers – not to buy products from forests converted after that date.

These observations do not mean that either the palm oil sector or users of palm oil products are justified in making broad-brush statements that ignore facts, or refuse to address specific points of detail, or absolve them of responsibility if species go extinct. By the same token, commentators who write reports or produce documentaries for global distribution should not cherry-pick dramatic sound-bites or images to imply that there is a simple linear relationship between cutting down trees, expansion of oil palm plantations and the threatened status of orang-utans.

HOW NON-GOVERNMENTAL ORGANIZATIONS CAN HELP

NGOs generally already know what they want to achieve. Those that work on the ground know that they need to choose and maintain focus. They cannot stop irreversible global trends. The best way is to work on one or two achievable goals. That might be saving and rehabilitating displaced orang-utans or working to help the authorities identify, create and sustain a new protected area.

HOW ENTREPRENEURS CAN HELP

By definition, entrepreneurs should have their own creative ideas, but here are two for a start.

Global warming is driven in large part by the increasing concentration of carbon dioxide in the air. A part of the way in which carbon dioxide emissions can be mitigated is to absorb carbon dioxide on a large scale. One way to do that is to plant trees on a large scale, choosing species that will grow to a large size, capturing large volumes of carbon dioxide per hectare. Funds to do the planting could be obtained from or by companies or institutions that wish to invest in 'carbon credits' that could later be traded as part of national or regional schemes for carbon trading. There are already precedents. It is now a matter of seeking the investors and arranging the deals – in this case to restore damaged orang-utan habitats.

Imagine creative new ways to include wild orang-utans within new nature tourism products. For example, it might be possible to view the orang-utans from tree walkways or even balloons.

HOW INDIVIDUALS CAN HELP

The combined forces of human population growth, poverty, unpredictable El Niño droughts, and growing international trade in commodities such as palm oil, rubber and wood products are, to a large degree, too strong to be influenced greatly by individual concerns over a single species, even a very special one such as the orang-utan. The frustration generated by that harsh fact has led to the production of several documents and documentaries in recent years which lash at the most prominent, or most immediate, perceived culprit in the demise of the orang-utan, namely oil palm plantations.

Concerned individuals, governments (at local, regional, national and international levels), non-governmental organizations, industries (notably palm oil and wood) all have roles to play to ensure that orang-utans survive. Repeated constructive engagement, again and again, may achieve results. Polarization of positions will not.

Individuals who decide seriously to play a role in helping orang-utans would first do well to try to identify as specifically as possible what aspect he or she is possibly able to influence and is most interested in supporting. One way to start is to decide whether you are primarily concerned about the welfare of individual displaced orang-utans or about the retention and rehabilitation of natural orang-utan habitats. The next is probably to decide what you are able to do.

The simplest way to help might be to write letters to key people or institutions. This requires some careful research on whom one might usefully write to, what to say, how to say it and how to ensure that the letter will reach its target. Letters may not be of much value in helping displaced orang-utans, because the cause of their displacement may already be entrenched. However, letters can play a role in some specific cases, such as pressing for captive orang-utans, held under bad conditions outside their country of origin, to be returned to an established rehabilitation centre. Letters can be of potential value in alerting government ministers, or other high-level office-bearers in government agencies, or companies with large land holdings in Borneo or Sumatra, to the identity and significance of specific orang-utan habitats that merit attention. Advice needs to be obtained – perhaps from established non-governmental organizations operating in regions where orang-utans occur – on which issues would be helped by letters from concerned strangers. In any case, there are three golden rules. First, one needs to acquire a clear understanding of all aspects of the issue. Paragraphs gleaned from international press releases or the Internet may be a gross simplification, or be out of date. Second, the letter must be clear, as short as possible, and polite. And third, the letter must be addressed and delivered to the person who is in a position to take action.

Another, even simpler but more costly option is to donate money to established non-governmental organizations or groups working on orang-utan rehabilitation projects, or habitat conservation projects, or good quality research related to orang-

utans. Use of an Internet search engine can track down most, and probably all, such organizations and projects. The task is then to choose some attractive ones, review them, and decide which to support. It is worth remembering that, while all conservation projects want more money, an even bigger problem may be consistency of funds. It is better for orang-utan conservation projects to receive a moderate and consistent stream of money than to get lots of funds followed by nothing.

Opposite *Oil palm plantations yield an average of over four tonnes of orange-coloured edible oil per hectare per year, far more than any other crop plant.*

Above *Orang-utans can often be seen from a boat on the scenic Kinabatangan River in Sabah – surely other appropriate ways exist for wild orang-utans or their habitats to be an 'ecotourism product'?*

More complicated, and possibly risky, would be to try to join one of the existing field projects as a volunteer or staff member: complicated, in terms of taking time off for extended periods, along with financing, immigration and lifestyle issues, and risky in terms of health, mental adjustments, or disruption of career. An option for those with a university degree in a biological or social subject, chosen by several well-known orang-utan experts, might be to find a university and funds in order to pursue a higher degree on wild orang-utans.

It may be necessary to decide whether to focus your interest on Indonesia or Malaysia, as there are big differences between the two, in terms of how to best target help for orang-utans. For orang-utan rehabilitation in Malaysia, the target would probably be the Sepilok orang-utan rehabilitation centre, which has been operating since 1965 and is managed by the Sabah Wildlife Department. In Indonesia, the era of orang-utan habitat loss, and of displaced captive orang-utans, is far from over in both Kalimantan and Sumatra, so there will be a need for financial support for years to come. There are several choices of project, mostly run or supported by non-governmental organizations. Some background research will be needed to get a feel for them all. If your interest is orang-utan habitat in Malaysia, then a good option might be to help raise funds towards the rehabilitation of damaged habitat. This is because the pattern of plantations and permanent forest cover where wild orang-utans occur is now stabilizing, but several important habitats in Sabah have been damaged by a long history of logging and, in some areas, fire during droughts. Specific areas in Sabah where important orang-utan habitat rehabilitation efforts are already underway include Ulu Segama Malua Forest Reserves (representing the extensive logged forest surrounding Danum Valley Conservation Area) and the lower Kinabatangan floodplain, where there is a mix of Wildlife Sanctuary, Forest Reserve and private land. Agencies working on forest rehabilitation efforts in these areas include the Sabah Forestry Department, Sabah Foundation, WWF-Malaysia, Kinabatangan Orang-utan Conservation Project-Hutan and several smaller institutions based in local villages. These efforts will require large amounts of sustained funding over many years to come. Although small individual donations can always help, significant sources of long-term funding may come in the form of voluntary carbon dioxide absorption projects.

For those who have the time and inclination for wide and regular reading and analysis of global trends in the oil palm and wood-based industries, as well as orang-utans and tropical forests, it is possible to develop a long-term interest. By tracking these global issues and trends, new and unexpected opportunities to help orang-utans may become apparent, not always obvious to others who are close to the ground. There may be opportunities for lobbying, or working with international corporations, or noticing that one old threat is now subsiding and that it will be better to devote conservation effort towards a new emerging issue.

One example comes in the British government's 2006 Stern Review on the economics of climate change, which advocates that developed nations should pay developing nations to reduce deforestation, and that the opportunity costs of not converting forests for alternative economic uses should be born by the international community. To avoid the risk that such a scheme might be seen as infringing the rights of Indonesia and Malaysia to develop, it must be clear that these nations are providing a globally significant service – saving orang-utans along with a lot of other biodiversity. They are not just seeking handouts. And in return, the recipient nations will have to be clear and committed on their responsibilities.

Finally, one might consider a vacation built around visiting one or more wild orang-utan habitats in Malaysia or Indonesia. This is because, whether we like it or not, governments of both Malaysia and Indonesia are keen to promote tourism long-term. For that reason, and as long as international tourists come to these countries in part to see wildlife, governments will tend to protect habitats where there are popular wildlife species such as orang-utans. Well-established tours to readily accessible orang-utan habitats are available in Sabah, Sarawak and also Sumatra, for those with moderate to higher budgets. In Sabah, low-budget options with a very good chance of seeing a wild orang-utan also exist. Visits to orang-utan habitats in Kalimantan and other parts of Sumatra require more time, effort and a rather higher budget to cover the extra time and travel costs.

Opposite (top) *The Nyaru-Menteng orang-utan rescue and rehabilitation centre is of special significance in helping to save the numerous orang-utans still being affected by oil palm plantation expansion and forest fires in Central Kalimantan.*

Opposite (bottom) *Tourism involving orang-utans helps to ensure that their habitats will be conserved; Kota Kinabalu, the capital of Sabah, is the major gateway and new opportunities may emerge elsewhere in Borneo and Sumatra in the future.*

BOS-WANARISET

ORANGUTAN REINTRODUCTION CENTER

Established in 1991, BOS-Wanariset in Samboja, East Kalimantan is the first Orangutan Reintroduction Center managed by BOS. At BOS-Wanariset, ex-captive orangutans go through all stages of Orangutan Reintroduction Program: quarantine, socialization, pre release and release.

As one of Indonesia's Animal Rescue Centers, BOS-Wanariset currently cares for more than 200 orangutans, more than 30 sun bears around 10 birds. Since its beginning, the Center has been supporting the Department of Forestry in conducting extensive confiscation activities, has cared for more than 1000 orangutans and released around 400 of them to the wild.

Didirikan pada tahun 1991, BOS-Wanariset yang terletak di Samboja, Kalimantan Timur adalah Pusat Reproduksi Orangutan pertama yang dikelola oleh Yayasan BOS. Di BOS-Wanariset, orangutan hasil penyelamatan atau sitaan melalui semua tahap-tahap Program Reintroduksi Orangutan: karantina, sosialisasi, pre pelepasliaran dan pelepasliaran.

Sebagai salah satu Pusat Penyelamatan Satwa di Indonesia, BOS-Wanariset saat ini merawat lebih dari 200 orangutan, lebih dari 30 beruang madu dan sekitar 10 ekor burung. Sejak awal, Pusat Reintroduksi ini telah mendukung Departemen Kehutanan banyak kegiatan penyelamatan dan penyitaan satwa liar yang dilindungi, merawat lebih dari 1000 orangutan dan telah melakukan pelepasliaran 400 orangutan ke habitat alam mereka.

WHERE TO SEE WILD ORANG-UTANS

Sabah

Danum Valley Conservation Area (438 sq km) and the surrounding Ulu Segama Malua Forest Reserves (2,400 sq km)

This complex supports the largest Malaysian population of orang-utans, with an estimated 4,300 individuals, most in the logged lowlands surrounding Danum Valley. Specialist nature tours are available with a good chance to see wild orang-utans.

Deramakot Forest Reserve (550 sq km)

It is estimated that this timber production area, certified by the Forest Stewardship Council, supports about 770 wild orang-utans.

Kinabatangan Wildlife Sanctuary and adjacent Forest Reserves (39,400 sq km)

Most of this area was allocated by government policy in the 1970s for conversion to agriculture, as part of a vision to diversify Sabah's economy. As a consequence, there was no restriction on logging. Following a state-wide survey of the distribution of orang-utans by WWF-Malaysia with the Sabah Forestry Department in the mid-1980s, it was realized that the Kinabatangan River floodplain supported high population densities of orang-utans, despite the poor condition of the remaining forest. Kinabatangan Orang-utan Conservation Project has been studying and monitoring the orang-utans in this area since 1998. An estimated 1,100 orang-utans live in the area.

A variety of general and specialist operators offer tours to lower Kinabatangan, with one offering guaranteed sightings of wild orang-utans.

Kulamba Wildlife Reserve (204 sq km)

A coastal area consisting of a mix of freshwater, transitional dry land and beach forest, unique in being almost undisturbed by human presence, and containing about 900 orang-utans. Although only two hours by boat from Sandakan, it is rarely visited because of the absence of a reliable freshwater supply for human visitors and its inhospitable, swampy character.

Sepilok Forest Reserve (43 sq km)

Established as a protected area in the 1950s, and chosen as a rehabilitation centre for young orang-utans in 1964. Easy road access from Sandakan since the 1970s has made Sepilok a prime location for visitors wishing to see orang-utans in a forest setting. There is an almost guaranteed chance to see young orang-utans at feeding times, and a rainforest discovery centre nearby.

Tabin Wildlife Reserve (1,225 sq km)

Established in 1984 as a refuge for Sumatran rhinoceros and Borneo pygmy elephant, Tabin has become an important Sabah site for translocation of orang-utans displaced by forest conversion elsewhere. Specialist nature tours are available, with a possibility of seeing orang-utans.

Sarawak

Batang Ai National Park (240 sq km) and adjacent Lanjak-Entimau Wildlife Sanctuary (1,688 sq km)

Lowland and hill dipterocarp forest and old secondary forest. The National Park, 250km from Kuching, with resort hotel, lodge, camp and Iban longhouse accommodation, forms the catchment of a hydro-electricity complex. Trails exist but spotting orang-utans is not guaranteed. Visits are normally four days, including boat trips and visits to members of the local Iban community.

West Kalimantan

Betung Kerihun National Park (8,000 sq km)

Contiguous with Lanjak-Entimau in Sarawak, this is an important area for conservation of the orang-utan. Access is via Putussibau and suitable for the adventurous visitor.

Gunung Palung National Park (900 sq km)

A unique mix of several forest types, and the location of many research projects on wild orang-utans since 1985. Trails exist and there is a good chance of seeing orang-utans. Visits normally involve a journey of several days, starting from Pontianak via Ketapang, and incorporating local community aspects.

Central Kalimantan

Nyaru-Menteng (0.6 sq km)

Now the most active orang-utan rehabilitation and reintroduction project in Indonesia, situated 28km by road from Palangkaraya.

Sebangau National Park (5,687 sq km)

Has a very large orang-utan population in peat swamp forests. Access via Palangkaraya.

Tanjung Puting National Park (3,040 sq km)

A well-established park, with swamp forest habitats and a large orang-utan population. Access via Pangkalan Bun.

East Kalimantan

Kutai National Park (1,986 sq km)

Most of the Park has been severely damaged by fire and encroachment, but wild orang-utans can be seen. Access is from Samarinda to Mentoko via the Bontang–Sangatta road.

Sumatra

Gunung Leuser National Park (9,500 sq km)

The easiest access and chance to see orang-utans is at Bukit Lawang (also known as Bohorok), 80km by road from Medan.

Opposite (left) *River tourism, Kinabatangan Wildlife Sanctuary.*

Opposite (right) *Bukit Lawang, Gunung Leuser National Park.*

Above *The most convenient way for people to see orang-utans in a natural forest environment is at any of the orang-utan rehabilitation centres in Indonesia or Malaysia.*

FURTHER READING

This list provides only examples of the wide range of publications on orang-utans, available either through bookshops or the Internet.

Ancrenaz, M. (2006) 'Consultancy on survey design and data analysis at Betung-Kerihun National Park, Indonesia'. Report to WWF-Germany. (see **http://www.wwf.or.id/admin/file-uploads/files/FCT1165193058.pdf**)

Ancrenaz, M., Giminez, O., Ambu, L., Ancrenaz, K., Andau, P., Goossens, B., Payne, J., Tuuga, A., Lackman-Ancrenaz, I. (2005) 'Aerial surveys give new estimates for orang-utans in Sabah, Malaysia'. *PloS Biology* 3(1):e3. (see http://dx.doi.org/10.1371/journal.pbio.0030003)

Caldecott, J., Miles, L, eds (2005) 'World Atlas of Great Apes and their Conservation'. Prepared at the UNEP World Conservation Monitoring Centre. University of California Press, Berkeley, USA.

Friends of the Earth (2005) 'The Oil for Ape Scandal – How palm oil is threatening the orang-utan'. (see **http://www.foe.co.uk/resource/reports/oil_for_ape_summary.pdf**)

Kilbourn, A., Karesh, W., Wolfe, N. D., Bosi, E. J., Cook, R. & Andau, M. (2003) 'Health evaluation of free-ranging and semi-captive orangutans (Pongo pygmaues pygmaeus) in Sabah, Malaysia'. *Journal of Wildlife Diseases* 39(1):73-83.

MacKinnon, K.S., Hatta, G., Halim, H., Mangalik, A. (1996) *The Ecology of Kalimantan*. Periplus, Singapore.

Nelleman, C., Miles, L., Kaltenborn, B.P., Virtue, M., Ahlenius, H. (eds.) (2007) 'The last stand of the orangutan – state of emergency: Illegal logging, fire and palm oil in Indonesia's national parks'. United Nations Environment Programme, GRID-Arendal, Norway.

Rijksen, H., Meijaard, E. (1999) *Our Vanishing Relative: The Status of Wild Orang-utans at the Close of the Twentieth Century*. Kluwer Academic Publishers, Dordrecht.

Schaik, C. van (2004) *Among Orang-utans: Red Apes and the Rise of Human Culture*. Harvard University Press.

Singleton, I., Wich, S., Husson, S., Stephens, S., Utami Atmoko, S., Leighton, M., Rosen, N., Traylor-Holzer, K., Lacy, R., Byers, O., eds (2004) 'Orangutan Population and Habitat Viability Assessment: Final Report'. IUCN/SSC Conservation Breeding Specialist Group, Apple Valley, Minnesota. (see **http://www.cbsg.org Reports/PHVA Reports/Orangutan PHVA 15-18 January 2004**)

Whyte, S., Desilits, M., Warwick, H. (2006) 'Save orangutans from extinction when you next shop'. (A joint publication of Nature Alert and the Borneo Orangutan Survival Foundation UK). (see **http://www.orangutan.org.au/assets/images/palmoil/BOSPalm_Oil_Report.pdf**)

Whitten, A.J., Damanik, S.J., Anwar, J., Hisyam, N. (1984) *The ecology of Sumatra*. Gadjah Mada University Press, Yogyakarta, Indonesia. Reprinted (2000), Periplus Singapore.

Wolfe, N.D., Kilbourn, A.M., Karesh, W.B., Hasan, A.R., Bosi, E.J., Cropp, B.C., Andau, M., Spielman, A., Gubler, D.J (2001) 'Sylvatic transmission of arboviruses among Bornean orangutans'. *Am. J. Trop. Med. Hyg.* 64(5&6):310-316 (see http://www.ajtmh.org/cgi/reprints/64/5/310/pdf)

Weblinks

The weblinks listed below represent the wide range of organizations concerned with orang-utans and their conservation; many more can be found via the Internet.

The Great Apes Survival Project (GRASP) Partnership is a project of the United Nations Environment Programme (UNEP) and the United Nations Educational, Scientific and Cultural Organization (UNESCO), which aims to lift the threat of extinction faced by all the living non-human great apes (**www.unep.org/grasp**).

WWF, the global environmental conservation organization, raises funds and conducts projects towards orang-utan conservation (**www.panda.org**; see also **www.wwf.de** (Germany), **www.wwf.or.id** (Indonesia), **www.wwf.org.my** (Malaysia), **www.wwf.nl** (Netherlands), **www.wwf.org.uk** (UK) and **www.worldwildlife.org** (USA)).

A range of non-governmental organizations support local orang-utan conservation efforts in Indonesia and Malaysia, including Borneo Orangutan Survival Foundation UK (**www.savetheorangutan.org.uk** and **www.savetheorangutan.co.uk**), Orangutan Conservancy (**www.orangutan.com**), Orangutan Foundation UK (**www.orangutan.org.uk**), Orangutan Foundation International (**www.orangutan.org**) and Sepilok Orangutan Appeal UK (**www.orangutan-appeal.org.uk**). The Orang-utan Tropical Peatland Project is concerned with long-term research in the Sebangau peat swamp forests (**www.orangutantrop.com**), Red Ape Encounters provides community-based orang-utan ecotours (**www.redapeencounters.com**) and The Sumatran Orangutan Conservation Programme conducts a variety of activities to help save the Sumatran orangutan (**www.sumatranorangutan.org**).

Born To Be Wild focuses on the trade, abuse and treatment of orang-utans throughout the world (**www.born-to-be-wild.org**).

The Environmental Investigation Agency (EIA) is an international campaigning organisation that investigates environmental crimes (**www.eia-international.org**).

The Roundtable on Sustainable Palm Oil (**www.rspo.org**) is an association created by organisations carrying out their activities in and around the entire supply chain for palm oil, aiming to promote the growth and use of sustainable palm oil through co-operation and open dialogue amongst stakeholders.

Recordings of some of the most common wild orang-utan vocalizations can be heard at: **www.aim.uzh.ch/orangutannetwork/Orangutancallrepetoires.html#6**

INDEX

ACKNOWLEDGEMENTS

The author and photographer wish to thank the following for their support, assistance and provision of information that contributed towards the production of the text and photographs in this book: Dr Dionysius S. K. Sharma (Executive Director/CEO), Dr Rahimatsah Amat, Mr Raymond Alfred, Ms Lee Shan Khee, Mr Darrel Webber, and staff of WWF-Malaysia; Mr Mahedi Andau (Director), Mr Laurentius Ambu, Mr Augustine Tuuga, Dr Sen Nathan, Dr Cecilia Boklin, Ms Sylvia Alsisto and the rangers and staff of Sabah Wildlife Department; Datuk Sam Mannan (Director), Mr Frederick Kugan and staff of Sabah Forestry Department; Dr Waidi Sinun, Mr Jimmy Omar and staff of the Yayasan Sabah Group; the Chief Minister's Department of the Government of Sabah; Dr Marc Ancrenaz, Dr Isabelle Lackman-Ancrenaz and staff of Kinabatangan Orang-utan Conservation Project; Carol Angkangon Prudente and the staff of North Borneo Safari Sdn. Bhd.; Dr Melvin Gumal, Director of WCS-Malaysia Program (Sarawak); Dr Mubariq Ahmad (Executive Director/ CEO) and staff of WWF-Indonesia; Ms Lone Dröscher-Nielsen (Manager) and Mr Hardi Baktiantoro (Vice Manager), BOS Nyaru Menteng; Cecep Munaiat, Orangutan Foundation International, Camp Leakey, Tanjung Puting National Park; Dr Birute Galdikas, Orangutan Foundation International; Herry Roustaman, Borneo Wilderness, Kumai, Central Kalimantan; Dr Sri Suci Utami Atmoko, Universitas Nasional, Jakarta, Indonesia; Dr Ian Singleton, Sumatran Orangutan Conservation Project; Mr Politarius, Indonesia Ecoventure; Mr Stefan Ziegler and staff of WWF-Germany; Mr Roland Melisch, TRAFFIC International; Wendy Elliott, Diane Walkington and staff of WWF-UK; Wim Ellenbroek and staff of WWF-Netherlands; Dr Christian R. Schmidt, Director of Frankfurt Zoo; Dr Stephan Hering-Hagenbeck, Director, Tierpark Hagenbeck Hamburg; Amanda Embury, Primate Taxon Advisory Group, Australasian Species Management Program of the Australasian Regional Association of Zoological Parks and Aquaria; Dr Clemens Becker, European Endangered Species Programme for Orang-utans, European Association of Zoos and Aquaria; Dr Lori Perkins, Orang-utan Species Survival Plan, Association of Zoos and Aquariums, USA; and Mr Quentin Phillipps.

In addition to the people and organizations named above, the contents of this book owe much to the efforts and research of many others. Special mention goes to Dr John MacKinnon, Dr Herman Rijksen, Dr Carel P. van Schaik, Dr Cheryl Knott, Dr Helen Morrogh-Bernard, Dr Erik Meijaard and Dr Benoît Goossens.

The opinions expressed in this book do not necessarily reflect those of any scientist, individual or organization mentioned in the book.